THE SAN JOAQUIN KIT FOX

THE SAN JOAQUIN KIT FOX

Biology, Ecology, and Conservation of an Endangered Species

BRIAN L. CYPHER

With a Foreword by Claudio Sillero-Zubiri

Comstock Publishing Associates
an imprint of
Cornell University Press

Ithaca and London

Publication of this book was supported with principal investigator discretionary funds from the California State University-Stanislaus.

First published 2024 by Cornell University Press

Printed in China

Library of Congress Cataloging-in-Publication Data
Names: Cypher, Brian L., author. | Sillero-Zubiri, Claudio, writer of foreword.
Title: The San Joaquin kit fox : biology, ecology, and conservation of an endangered species / Brian L. Cypher, with a foreword by Claudio Sillero-Zubiri.
Description: Ithaca : Comstock Publishing Associates, an imprint of Cornell University Press, 2024. |
Includes bibliographical references and index.
Identifiers: LCCN 2024008961 (print) | LCCN 2024008962 (ebook) | ISBN 9781501775055 (hardcover) | ISBN 9781501775079 (epub) | ISBN 9781501775062 (pdf)
Subjects: LCSH: San Joaquin kit fox. | San Joaquin kit fox—Ecology. | San Joaquin kit fox—Conservation.
Classification: LCC QL737.C22 C97 2024 (print) | LCC QL737.C22 (ebook) | DDC 599.775—dc23/eng/20240324
LC record available at https://lccn.loc.gov/2024008961
LC ebook record available at https://lccn.loc.gov/2024008962

*To Jerry H. Scrivner (1952–2020),
a knowledgeable, kind, and gentle
soul, and without whose patience
and guidance I might not have
become a kit fox biologist.*

*My wonderful partner, Ellen, has
supported me throughout our life
together, and this support has been
the foundation for all happiness and
any success I have had.*

*Finally, this book is dedicated to
San Joaquin kit foxes. They have taught
me much and have been the basis
of a fine career. I hope to return
those favors in some small measure
through this book.*

CONTENTS

List of Illustrations   viii
Foreword by Claudio Sillero-Zubiri   xi
Preface   xv
Acknowledgments   xix

Introduction:
Why a Book About the San Joaquin Kit Fox?   1

1 *San Joaquin Kit Fox Overview*   5
Description   5
Adaptations   11
Ecology   12
Status   14
Evolution and Taxonomy   16

2 *Natural History*   21
Distribution and Habitat Preferences   21
Survival and Mortality Factors   41
Reproduction   60
Population Dynamics   64
Foraging Ecology   73
Space Use, Movements, Dispersal, and Activity   79
Dens   89
Social Ecology   102
Behavior   108
Interspecific Interactions   113

3 *Contemporary Challenges*   119
   Kit Foxes and Oil Fields   119
   Kit Foxes and Solar Farms   123
   Kit Foxes and Agricultural Lands   131
   Kit Foxes and Roads   136
   Kit Foxes and Urban Areas   140
   Kit Foxes and Climate Change   158

4 *Conservation*   163
   History of Research Efforts   163
   History of Conservation Efforts   168
   Biopolitics and Attitudes   176
   Research and Conservation Needs   183

Conclusion: *The Future*   189

Appendix: Common and Scientific Names of Species   195
Literature Cited   199
Index   217

ILLUSTRATIONS

1.1. San Joaquin kit foxes with winter and summer pelage. 6
1.2. Other canids confused with kit foxes. 8
1.3. San Joaquin kit fox at a den. 12
1.4. Range map for kit foxes. 15

2.1. Map of San Joaquin Desert in central California. 25
2.2. Map of core populations of San Joaquin kit foxes. 27
2.3. Map of satellite populations of San Joaquin kit foxes. 29
2.4. Map of suitable habitat for San Joaquin kit foxes. 36
2.5. Highly suitable habitat for San Joaquin kit foxes. 37
2.6. Low suitability habitat for San Joaquin kit foxes. 38
2.7. San Joaquin kit fox killed by a coyote. 46
2.8. San Joaquin kit fox killed by a bobcat. 46
2.9. San Joaquin kit fox killed by a golden eagle. 47
2.10. San Joaquin kit fox killed by a vehicle. 49
2.11. San Joaquin kit fox with flea infestation. 52
2.12. San Joaquin kit foxes with sarcoptic mange. 55
2.13. Kit foxes surveyed in Bakersfield, CA, 2015–2022. 56
2.14. A San Joaquin kit fox before and after treatment for mange. 57
2.15. San Joaquin kit fox bringing kangaroo rats to a natal den. 61
2.16. San Joaquin kit fox pups 4–5 weeks old. 62
2.17. San Joaquin kit fox pups about 10–12 weeks old. 63
2.18. Three kangaroo rat species consumed by San Joaquin kit foxes. 68

2.19. Number of kit foxes relative to a precipitation model.   70

2.20. Giant kangaroo rat.   74

2.21. Giant kangaroo rat precincts.   75

2.22. San Joaquin kit fox home range overlap on two study sites.   84

2.23. "Keyhole"-shaped den entrance.   92

2.24. Long "berms" common outside of kit fox dens.   93

2.25. Natal den of a San Joaquin kit fox.   95

2.26. Pup scats and kangaroo rat remains outside of natal dens.   96

2.27. Artificial den designs and materials tested for kit fox preferences.   98

2.28. Kit foxes using artificial dens.   99

2.29. Recommended artificial den design and materials.   100

2.30. Adult kit foxes and a female helper attending to pups.   105

2.31. Two female kit foxes that have both reproduced.   107

2.32. A San Joaquin kit fox fecal latrine.   111

2.33. Kit fox fecal deposit on a small mammal trap.   111

2.34. A red fox entering a San Joaquin kit fox den.   116

2.35. A striped skunk entering a San Joaquin kit fox den.   118

3.1. San Joaquin kit fox covered in oil.   121

3.2. Fences around solar facilities modified for passage by kit foxes.   125

3.3. Kit foxes in construction materials and under a construction vehicle.   128

3.4. San Joaquin kit fox under solar panels.   129

3.5. Agricultural lands in the San Joaquin Valley, California.   132

3.6. San Joaquin kit foxes on grazing land.   134

3.7. San Joaquin kit fox killed by a vehicle.   137

3.8. Remains of a kit fox killed by a vehicle near median barriers.   139

3.9. San Joaquin kit fox in Bakersfield.   141

3.10. San Joaquin kit foxes caught in sports nets.   144

3.11. Litters of San Joaquin kit fox pups.   146

3.12. San Joaquin kit foxes with food items in Bakersfield.   147

3.13. San Joaquin kit fox dens in Bakersfield.   150

3.14. San Joaquin kit fox dens in inconvenient locations.   155

3.15. Map of potential range expansion by San Joaquin kit foxes under climate change.   160

4.1. Recovery Plan for Upland Species of the San Joaquin Valley, California.   172

4.2. San Joaquin kit fox at the Carrizo Plain National Monument.   174

4.3. *After the Grizzly* by Peter Alagona.   177

4.4. San Joaquin kit foxes on college campuses.   179

4.5. Ambassador San Joaquin kit fox at the California Living Museum.   188

C.1. Litter of San Joaquin kit fox pups.   192

urban uses and crisscrossed by oil and linear infrastructure. Countless foxes succumbed to strychnine campaigns targeting coyotes. Many of these threats remain today, and additional contemporary challenges result from the advance of solar farms, expanding road networks, and urban sprawl. Intriguingly, we cannot anticipate how climate change might play out, and it might well work to the foxes' advantage, since it is quite possible that the land set aside for agriculture might shrink and naturally revert, or it might be restored and become more suitable for kit foxes and other wildlife.

This is a wonderful book, a summary of what we know about San Joaquin kit foxes, and it will satisfy different users, from the amateur enthusiast trying to learn more about local wildlife, to the scholar mining information for a comparative study. While a vast body of literature has been published on these foxes, there is enormous value in a synthesis as a guide for ongoing and future conservation efforts for this endangered species. The wealth of information crammed into this text should be of use to those planning and implementing relevant conservation actions, and it will help in the survival of these intriguing foxes as they navigate the challenges of life in a highly intervened landscape.

Claudio Sillero-Zubiri
Professor of Conservation Biology
Wildlife Conservation Research Unit
Department of Biology, University of Oxford
Chair, IUCN SSC Canid Specialist Group

PREFACE

The preparation of this book in large part represents a personal journey driven by a sense of responsibility, as I'll attempt to explain. I first began working with San Joaquin kit foxes in July 1990, and that work continues to my writing of this book (and hopefully beyond!). I did not specifically set out during my early schooling and training to become a kit fox biologist, although it was certainly not outside the realm of possibilities. I have always had a deep fascination and interest in carnivores, particularly wild canids. But as with many in the field of wildlife ecology in the late 1970s–early 1980s, I mostly just wanted a career working with wildlife. At the State University of New York, College of Environmental Science and Forestry in Syracuse, among other endeavors I assisted my undergraduate adviser with a food habit study on coyotes and red foxes in the Adirondack Mountains. This helped pique an interest in predator-prey relationships. After completing my Bachelor of Science degree and prior to entering a Master of Science program, I took additional classes at the University of Minnesota (where my wife, Ellen, was working on her MS degree) and then had a wonderful opportunity to work on several aspects of the gray wolf project being conducted by Dave Mech and Uly Seal. This was a dream come true for someone at such an early stage in their career. I learned yet more research techniques, got additional field experience, and also began turning an eye more toward conservation of species instead of management of game populations. My MS research at The Pennsylvania State University addressed white-tailed deer ecology and management at Valley Forge National Historical Park; however, I also managed to sneak in a small study on red fox food habits in the park.

My PhD project addressed the ecology of coyotes at the Crab Orchard National Wildlife Refuge in Illinois. In the winter and spring of 1990, I was working on completing my dissertation at Southern Illinois University. I was not thinking too seriously yet about jobs or what might come next. In the back of my head, I was toying with maybe conducting a postdoctoral project somewhere, maybe even some place exotic like Africa. Ultimately, I was hoping to land in a faculty position at a university. Neither of these two ventures ever came to pass, at least not in the way I expected or hoped they might. Although I was focused on wrapping up my dissertation, Ellen (who was in the final stages of her own PhD program in Botany) was checking the academic job ads to see what was out there. One day she showed me a job announcement in the back of *Science* magazine. Something about working with kit foxes and other endangered species as part of a conservation program being conducted by, of all things, a private company at a federally owned oil field in central California. The description did say something about endangered kit foxes, which certainly meshed with my now well-developed interest in wild canid conservation, but nothing else in the ad jumped out and grabbed me. Besides, I was still not quite done with my dissertation; however, Ellen persuaded me to apply. How fateful!

The company was EG&G Energy Measurements, Inc., a large tech company engaged in numerous Department of Energy and Department of Defense projects around the country, including evaluating the efficacy of nuclear devices at the Nevada Test Site; however, they also had a branch that conducted environmental programs at various locations. This group was born from the need for environmental compliance at the sites where the tech projects were conducted, but it also incorporated research and conservation whenever possible. The open position was for a person, preferably with some experience with wild canids, to participate in the environmental program run by EG&G at the Naval Petroleum Reserves in California (NPRC), consisting of Naval Petroleum Reserves No. 1 and 2, located about 30 miles west of Bakersfield in central California. (There also is an NPR-3 in Wyoming, the infamous "Teapot Dome," and an NPR-4 in Alaska.) These were indeed petroleum "reserves," the original purpose of which was to ensure that the United States Navy would always have a fuel supply. Then came the Arab oil embargo in 1973–1974, soon after which oil and gas production significantly increased at the Reserves. Then in 1977, the Department of Energy was created under the Carter administration to consolidate energy policy in the United States. The DOE soon assumed management of the Naval Petroleum Reserves.

At NPR-1 and NPR-2 (also known as the Elk Hills oil field and Buena Vista oil field, respectively), oil and gas production involved thousands of workers, thousands of wells, numerous processing and administrative facilities, and thousands of miles of pipelines and roads. Both facilities also encompassed considerable habitat, both in between the infrastructure and as large areas with little or no extraction activities. This habitat was inhabited by a number of officially Endangered as well as other rare species, including the giant kangaroo rat, Tipton kangaroo rat, blunt-nosed leopard lizard, San Joaquin antelope squirrel, Hoover's woolly-star, Kern mallow, and a robust population of San Joaquin kit foxes.

So I defended my dissertation at SIU on July 5, 1990, our belongings were packed into a moving van on July 10, we drove to California, and I started work with EG&G on July 15 with daytime temperatures topping 110 °F. In addition to the searing summer heat, a number of other things took some getting used to in Bakersfield (where we resided) and the San Joaquin Valley in general. An immense and immediate adjustment was transitioning from working in lush, green habitats to working in the arid scrub habitats occupied by San Joaquin kit foxes. These habitats officially qualify as "desert" habitats (see Germano et al. 2011), but they do not in any form resemble pretty deserts such as the Sonoran. Without getting into the details here, suffice it to say that Peter Alagona's description of the region in his book, *After the Grizzly* (Alagona 2013, 176), is spot-on. However, with time (and aided by the flushes of wildflowers in the spring in years with adequate rainfall, and also aided by the fact that the California coast is only two hours to the west and giant sequoias reside in the Sierra Nevada less than two hours to the northeast), one gains an appreciation for the area and learns how to deal with (or at least tolerate) its shortcomings.

Additionally, it helped that the community of species of which the San Joaquin kit fox is a member is truly fascinating and wonderful, and in dire need of help. Thus, in July of 1990, I began my long journey with San Joaquin kit foxes as well as with co-occurring species I have also had opportunities to get to study and know. I remained with EG&G and successor companies for 10 years. I had also been collaborating with the Endangered Species Recovery Program (ESRP), a research and conservation group associated with the California State University-Stanislaus. In 2000, I joined the staff of ESRP and therefore was able to continue working with San Joaquin kit foxes and many other species. As of this writing in 2023, I am still with ESRP and still conducting research and conservation projects.

When I came to the San Joaquin Valley in 1990, several other groups in addition to EG&G were also working on San Joaquin kit foxes, including Kathy Ralls and P. J. White with the Smithsonian Institution, Linda Spiegel and Dick Anderson with the California Energy Commission, and others. But their projects were of short duration and soon ended. Over the years, others have conducted short-term projects on San Joaquin kit foxes, but they have been relatively few in number and typically focused on narrow questions.

Thus many talented people have worked with the San Joaquin kit fox over time; however, for various reasons, including happenstance and in a "last man standing" sort of way, I have been fortunate in being able to conduct research and conservation efforts on San Joaquin kit foxes fairly continuously for more than 30 years. Although I hope to continue working with this critter in some capacity for some time, I acknowledge the reality of biology, and at the age of 64 I "feel my biological clock ticking" a bit. Thus, having accumulated 30-plus years of data, observations, experiences, and ponderings regarding the San Joaquin kit fox, I feel a responsibility to summarize and record this information, as well as incorporate the information collected by others, in a hope that it will benefit the species and also provide a base of knowledge on which other researchers and conservation biologists who follow can build. That is my purpose and goal for this book.

ACKNOWLEDGMENTS

I thank the many, many individuals with whom I have worked, collaborated, conversed, and sometimes agonized during my many years working with San Joaquin kit foxes. These individuals provided an abundance of information and perspectives that helped teach me, guide my efforts, and focus my thinking about San Joaquin kit foxes and their plight. Their collective input is included in this book and definitely improved it, and for that I am immensely grateful. I thank Scott Phillips and Tory Westall for assisting with the preparation of the map figures. I thank Erica Kelly, Nicole Deatherage, and Tory Westall for locating images from the vast ESRP photograph collection. Erica Kelly also assisted with the laborious task of checking citations in the text against the bibliography. I thank Ellen, my spouse, friend, and life-companion, for all her love and support before, during, and after the writing of this book. She also read the first draft and provided many helpful comments, suggestions, and edits. In addition, I thank valued colleagues Luke Hall and Tim Coonan, both of whom reviewed the first draft of the entire book and provided many positive comments and helpful suggestions. I also thank Claudio Sillero-Zubiri, another valued colleague, who graciously provided the foreword for the book. Finally, I thank Cornell University Press for agreeing to publish this book and for guiding me, a novice book author, through the process. I thank all the individuals at CUP who had any hand in the completion, production, and marketing of my book. In particular, I am especially grateful to Kitty Liu and Jacqulyn Teoh; their guidance, suggestions, and efforts were invaluable in completing the manuscript and shepherding it through the production process, and I am indebted to them. I also greatly appreciate the editing efforts of Susan Specter and Lucinda Treadwell that helped improve the writing.

THE SAN JOAQUIN KIT FOX

WHY A BOOK ABOUT THE SAN JOAQUIN KIT FOX?

So, why dedicate an entire book to the San Joaquin kit fox? Understandably, the reasons are unlikely to be clear to anyone not familiar with this animal. After all, the San Joaquin kit fox is not distinctively unusual in appearance or behavior. It does not have striking coloration like its close relative the red fox or have a familiar and universally recognized sound like the howl of a coyote. Being nocturnal and quite secretive, it is not a very obvious animal, although it can easily be seen if one knows where and how to look for it. Many people living and working in close proximity to kit foxes may never see one, although they may readily encounter its feces and dens. It is not economically important or culturally significant. The pelts are small and of relatively little value. Although it was certainly familiar to native peoples, it held no inordinately special place in their traditions and lore. Also, the San Joaquin kit fox is just a subspecies of a species that is relatively widespread and abundant. So, why does the San Joaquin kit fox warrant an entire book?

The San Joaquin kit fox is a predator, albeit a diminutive one, and predators have always been the recipients of considerable attention, both positive and negative (Kellert 1985; Bekoff 2001; Peek et al. 2012). Predators evoke strong emotions, and because of the adverse economic impacts they can cause, both real and perceived, they have frequently been reviled and persecuted. Although larger predators such as wolves, coyotes, mountain lions, and bobcats were usually the primary targets of predator control programs, kit foxes were frequent victims as well, either primary or collateral. Grinnell et al. (1937) reported that hundreds of San Joaquin kit foxes likely died each year in the early 1900s from strychnine campaigns targeting coyotes; likewise, they died from pest control programs, such as those that targeted ground squirrels on range lands (Schitosky 1975). Such programs undoubtedly had significant impacts on kit fox abundance, most certainly locally and possibly range-wide.

Furthermore, the range of the San Joaquin kit fox has always been limited to a relatively small geographic area, and therefore its numbers have concomitantly always been limited as well. At some point in the past, kit foxes from the Mojave Desert, where they are widespread and abundant, managed to cross the southern Sierra Nevada or Transverse Ranges and enter the southern Central Valley. Kit foxes are adapted to arid habitats, and therefore within the Central Valley, their range was largely limited to a 28,493 km^2 (11,000 mi^2) area defined as the San Joaquin Desert (Germano et al. 2011). (By comparison, the Mojave Desert is approximately 140,000 km^2 (54,000 mi^2) in size [MacMahon and Wagner 1985].) This range is bounded by high mountain ranges to the east and south and by more mesic and therefore less suitable habitats to the west and north. Beginning in the late nineteenth century, the amount of suitable habitat within this range shrank rapidly and profoundly due to conversion to agricultural, urban, and other uses. Consequently, the San Joaquin kit fox was included on the initial endangered species list issued in 1967 (USFWS 1998). Habitat loss in the San Joaquin Valley continues to this day, and the number of remaining San Joaquin kit foxes is estimated to be similar to the number of remaining giant pandas (Cypher et al. 2013).

In addition to being listed as federally endangered, the San Joaquin kit fox was eventually also listed as threatened under the California Endangered Species Act passed in 1972 (USFWS 1998). As a consequence of being both a federal and a state listed species, the San Joaquin kit fox has been the subject of a number of surveys and investigations; however, San Joaquin kit foxes have not enjoyed the notoriety or public interest enjoyed by other species such as California condors or sea otters or bald eagles. This lack of recognition is partly a consequence of the San Joaquin kit fox having the misfortune to occur in a region of California where economic wealth and environmental values are generally lower compared with other regions in California. Thus, resources for investigations or conservation efforts have been neither abundant nor consistently available. As a result, the San Joaquin kit fox has not benefited from long-term continuous study or population monitoring. Most efforts have been relatively short-term and have been conducted by a number of different individuals and organizations. This discontinuity in resources and efforts has made it extremely difficult to derive a comprehensive understanding of the ecology and conservation status of San Joaquin kit foxes. Even today, there is still a considerable lack of understanding of basic natural history attributes (e.g., optimal habitat characteristics, response to dynamic prey populations, and social ecology), including on the part of individuals and organizations responsible for overseeing the conservation and recovery of this species.

Because of ongoing societal tensions regarding endangered species and continuing development within the habitat of the San Joaquin kit foxes, they are occasionally at the center of conflicts and controversies. These situations are sometimes exacerbated by the gaps in general understanding, or sometimes by outright misconceptions, regarding kit foxes. For example, habitat that is marginal at best for kit foxes is sometimes treated as high quality in regulatory actions, and a common misconception is that kit foxes in inconvenient locations can simply be relocated elsewhere. Misconceptions and gaps in understanding can inhibit conservation and recovery efforts through ineffective and even harmful management decisions and erosion of public support.

Concomitant with the issues above, in recent decades the critical importance and value of predators has been recognized and increasingly embraced (Gittleman et al. 2001; del Rio et al. 2001; Peek et al. 2012; Bergstrom 2017). Their integral role in natural communities and ecosystem processes is unquestioned by ecologists and conservationists and is increasingly appreciated by the public at large. Examples of this evolution in attitude include the immense positive response to the reintroduction of gray wolves in Yellowstone National Park, visitation to Churchill, Manitoba, to view polar bears, greater tolerance of mesocarnivores in urban environments, and strong support for conserving endangered predators. Smaller species have not necessarily benefited from these changes in attitudes to the degree that larger predators have, but public interest in these species has grown as has a willingness to conserve them. Such interest as well as conservation efforts in general are facilitated and enhanced by the availability of current information.

For the many reasons detailed above, I feel that a book dedicated to the San Joaquin kit fox is warranted. Clearly, there is a need to summarize and synthesize the information on this species that currently is widely dispersed among numerous sources. Doing so will provide a more solid foundation for understanding this animal. Such a synthesis will also help inform management and conservation decisions, which in turn will help focus recovery efforts. Finally, it is hoped that such a synthesis will help foster a greater appreciation for the San Joaquin kit fox, particularly by the public whose support for conservation efforts is so critical. These are the objectives for this book. One additional note: much of the information presented is also applicable to other subspecies of kit foxes and even to closely related swift foxes.

This book is written in a semitechnical format that I hope will be easy for anyone with a basic biology background to follow and understand, as my intent is to further an understanding and appreciation for San Joaquin kit foxes. I did not want to write it in a completely popular format as I felt that too much important detail would be omitted. Nor did I want to write it in textbook format, as that would have made it relatively unavailable to those lacking a technical background. I strove to strike a balance. This book is generally organized into three major themes. In Chapters 1 and 2, biological and ecological information about the San Joaquin kit fox is summarized. In Chapter 3, details are provided on the contemporary challenges and potential threats facing the species. In Chapter 4, past research and conservation efforts are summarized, biopolitics are discussed, and suggestions for further research and conservation efforts are offered. I include some maps, graphs, and tables in an effort to make complex and detailed information more comprehensible and accessible. Images are distributed throughout the book to graphically illustrate important points. Scientific names are provided in an appendix to reduce clutter in the text and improve readability. I hope that readers will find all the material presented to be informative and interesting.

Chapter 1

SAN JOAQUIN KIT FOX OVERVIEW

To introduce the San Joaquin kit fox, a brief overview is provided, followed by a review of information on kit fox evolution and taxonomy. It is hoped this synopsis will be helpful to readers less familiar with this animal. The overview includes a description of the San Joaquin kit fox and summarizes its adaptations, ecology, and conservation status. For the most part, San Joaquin kit fox appearance, adaptations, and ecology are common to or at least very similar to those for all other kit foxes with a few exceptions. The taxonomy of the kit fox is still a topic of scientific inquiry and debate, although the implications for the San Joaquin kit fox, particularly its conservation, are minimal at best.

DESCRIPTION

The San Joaquin kit fox is a small canid. Indeed, the term "kit" is a testament to the small size of kit foxes, as this term is commonly reserved for very young animals (Grinnell et al. 1937). Like other kit foxes, the San Joaquin kit fox is roughly the size of a house cat or a Boston terrier. A thick coat in winter makes animals appear heavier, but thinner summer coats reveal the typical slender features (Figure 1.1). The pelage on kit foxes is grizzled to tawny gray, sometimes with buffy highlights on the neck, sides, and legs. The underside is commonly white to tawny, and the tail has a black tip. The feet have considerable hair between the pads, and the pinnae are lined with hairs, which presumably is an adaptation to exclude sand (McGrew 1979).

FIGURE 1.1. San Joaquin kit foxes with (a) winter and (b) summer pelage. Photos by Tory Westall.

Kit foxes exhibit a typical fox or vulpine body form with a narrow-pointed snout, large ears, and long bushy tail. Kit foxes are relatively delicate in appearance compared with other foxes. Most other diagnostic characteristics are typical of other foxes (Grinnell et al. 1937; McGrew 1979; Hall 1981). They have five toes on each forefoot (the fifth toe being a dew claw higher up on the side of the leg), four toes on each hind foot, and nonretractable claws. A subcaudal gland is present on the dorsal surface of the tail, and the hair over this gland may be darker.

Kit foxes overlap in distribution with other canids but are easily distinguished from most other species (McGrew 1979; Cypher 2003). Red foxes, per their name, have red pelage over most of their body, and they have dark coloration on their lower legs and the backs of their ears (Figure 1.2). They are two to three times the size of a kit fox and have a very distinctive, white-tipped tail. Gray foxes are about twice the size of a kit fox. They are stockier in appearance, with proportionally shorter legs and shorter ears. They usually have darker grizzled gray pelage, and they have a distinctive dark ridge down the dorsal surface of the tail. Adult coyotes are four to six times larger than a kit fox, and their pelage tends to have more variegated coloration. Adult coyotes are rarely confused with kit foxes because of their significantly larger size; however, young coyotes, particularly pups less than about 3 months old, are commonly confused with kit foxes, particularly in poor light or dense vegetation (O'Farrell 1987). This confusion has likely led to a number of putative kit fox observations that in reality were coyote pups (Clark et al. 2007).

Kit foxes and closely related swift foxes are very similar in appearance (McGrew 1979; Cypher 2003). However, the two species overlap only in a narrow (ca. 100 mi or 160 km) zone in eastern New Mexico and western Texas (Packard and Bowers 1970; Rohwer and Kilgore 1973). The range of the San Joaquin kit fox is far from that of swift foxes and is also disjunct from other subspecies of kit foxes (see Distribution and Habitat Preferences); thus, no other foxes similar in appearance are sympatric with San Joaquin kit foxes.

The San Joaquin kit fox is the largest of the extant subspecies of kit foxes. However, this difference would not be readily apparent to anyone who has not handled several individuals or at least observed several up close. Grinnell et al. (1937) reported the following mea-

A

B

FIGURE 1.2. Other canids that are sometimes confused with kit foxes including (a) a young coyote, (b) a red fox, (c) a gray fox, and (d) a kit fox for comparison. Photos by (a) "g'pa bill," (b) Dennis Jarvis, (c) James Marvin Phelps, and (d) Timothy Ludwig.

C

D

surements based on 10 male and 10 female specimens: For the males, mean total length was 805 mm (31.7 in), mean tail length was 295 mm (11.6 in), mean hind foot length was 124 mm (4.9 in), and mean external ear length was 87 mm (3.4 in). For the females, mean total length was 769 mm (30.3 in), mean tail length was 284 mm (11.2 in), mean hind foot length was 120 mm (4.7 in), and mean external ear length was 83 mm (3.3 in). In general, measurements for males are 2–7% greater than those for females (Grinnell et al. 1937). The dental formula of kit foxes is similar to that of other canids: incisors 3/3, canines 1/1, premolars 4/4, and molars 2/3. More detailed descriptions, including comparisons of external and skull morphology measurements between kit fox subspecies and between kit foxes and other species, can be found in Grinnell et al. (1937) and Hall (1981).

Mean mass (weight) of the specimens examined by Grinnell et al. (1937) was 2.2 kg (4.9 lb) for the males and 1.9 kg (4.2 lb) for the females. The lightest was 1.8 kg (4.0 lb) among the males and 1.5 kg (3.3 lb) among the females. The heaviest was 2.7 kg (6.0 lb) among the males and 2.3 kg (5.1 lb) among the females. Among the animals captured during years of research by the Endangered Species Recovery Program (unpublished data), the heaviest male was 3.1 kg (6.8 lb) and the heaviest female, 2.9 kg (6.4 lb).

As suggested by the numbers above, kit foxes exhibit mild sexual dimorphism in size, with males generally larger and heavier than females. Warrick and Cypher (1999) analyzed weights of 1077 known-aged kit foxes captured at the Naval Petroleum Reserves in western Kern County over 16 years (1980–1995). The foxes achieved about 90% of their adult weight by 10 months of age. Among adults, males were generally about 20% heavier than females. They also found that foxes were about 3.5% heavier in winter compared with summer. Finally, they found that mean weights of foxes varied annually, with foxes being heavier in years with higher prey abundance. Similarly, Spiegel (1996) reported that the mean weight for adults was closely correlated with annual precipitation, presumably because of greater prey abundance in years with higher rainfall. During the 16-year span of the Naval Petroleum Reserves study, there was about a 10% difference for males between the years with the heaviest and lightest weights and about a 12% difference for females (Warrick and Cypher 1999).

ADAPTATIONS

San Joaquin kit foxes exhibit a number of morphological, physiological, and behavioral adaptations associated with living in hot, arid environments. Many of these adaptations are designed to reduce heat loads and water loss as well as deal with the sandy conditions that typically occur in their habitats. Common among desert animals, kit foxes have relatively large ears that facilitate heat radiation. Dense long hairs along the inner edge and inner base of the pinnae cover the auricle opening and help exclude dust and sand (McGrew 1979). Kit foxes also have tufts of hair between their toe pads on the bottom of their feet, and this may improve traction on loose sand as well as protect the feet from the hot sand (Grinnell et al. 1937).

A significant challenge for desert animals including San Joaquin kit foxes is obtaining and retaining moisture. In an effort to lose excess heat, many animals pant, and this lowers body temperature through evaporative cooling, meaning that water is lost. Most species pant at one rate, that being the resonant frequency of the thorax. However, kit foxes are able to pant at a lower rate that increases as the ambient temperature increases until it reaches the resonant frequency. This results in a greater energy expenditure but reduces water loss (Denver Wildlife Research Center 1975 cited in McGrew 1979).

Kit foxes will drink free water if it is available (Egoscue 1956; personal observation). They have also been observed licking dew and frost off vegetation; however, kit foxes are able to obtain adequate moisture from their foods (Egoscue 1962; Morrell 1972). Metabolic water from the digestion and assimilation of prey provides approximately 18% of the total daily water requirement for kit foxes. The remainder of their requirement is met by preformed water from prey (Golightly and Ohmart 1984). To meet this requirement, kit foxes must consume substantially more prey than necessary to meet daily energetic demands. Golightly and Ohmart (1984) estimated that kit foxes need to consume 175 g (6 oz) of prey daily to meet their water requirements.

Despite living in hot environments, kit foxes are not tolerant of high ambient temperatures, and their body temperature can quickly rise to lethal levels at temperatures exceeding 35°C (Golightly and

FIGURE 1.3. San Joaquin kit fox at the entrance to an earthen den. Photo by Tory Westall.

Ohmart 1983). To conserve water, kit foxes rely minimally on evaporative heat loss and instead rely heavily on passive heat dissipation. Their bodies exhibit a high thermal conductance. Behavioral adaptations that help reduce heat loads include activity patterns that are primarily nocturnal. Also, kit foxes use earthen dens on a daily basis (Figure 1.3). During the day, kit foxes are down in dens where, during the hot weather, temperatures are lower and humidity is higher (see Dens).

ECOLOGY

Kit foxes are carnivores, and their diet consists almost exclusively of animal prey (see Foraging Ecology). They consume primarily rodents, with a particular emphasis on kangaroo rats, pocket mice, ground squirrels, and gophers. However, they also consume a considerable number of invertebrates, and in some years invertebrates may make up the majority of their diet. Plant material such as fleshy fruits are rarely consumed even when available. Food availability, particularly rodent abundance, appears to be the primary factor driving kit fox population dynamics (see Population Dynamics).

The primary cause of mortality for San Joaquin kit foxes is larger predators, particularly coyotes, bobcats, and golden eagles (see Survival and Mortality Factors). The next most common cause of mortality is vehicle strikes. Rodenticides can be an issue in very localized areas. Diseases generally tend not to be an issue for San Joaquin kit foxes, but there have been situations where diseases have had significant local effects.

The basic social unit for kit foxes is a mated pair (see Social Ecology). This pair is monogamous, although some extra-pair copulations do occur, and kit foxes commonly mate for life. The females are monstrous. San Joaquin kit foxes breed in early winter and give birth to a litter of pups in late winter (see Reproduction). Average litter size is four. Both members of the breeding pair provide parental care. The pups remain with the adults through the spring and then disperse in summer or fall. If food is abundant, one or more pups may delay dispersal and remain in their natal territory and may even assist their parents in raising the next litter of pups.

As mentioned previously, kit foxes are primarily nocturnal and only uncommonly observed during the day. They spend the day in dens, which are used on a year-round basis (see Dens). Each fox uses multiple dens per year, and these dens provide resting sites and escape cover, and they facilitate thermoregulation. Kit foxes occupy distinct home ranges (see Space Use, Movements, Dispersal, and Activity) but do not appear to be strongly territorial. Spacing among family groups appears to be maintained primarily through scent-marking.

San Joaquin kit foxes are geographically isolated from other kit foxes by the Sierra Nevada mountains (see Distribution and Habitat Preferences). They are restricted in distribution to the San Joaquin Desert region in central California, which includes the arid portions of the San Joaquin Valley and some adjacent smaller valleys. Within their range, they occur in arid shrub scrub and grassland habitats. These habitats have been significantly reduced through conversion to agricultural, urban, and industrial uses, resulting in profound reductions in the range and number of San Joaquin kit foxes. Consequently, they have been listed as federally endangered since 1967 and state threatened since 1971. San Joaquin kit fox range and numbers are still declining.

STATUS

As a species, kit foxes occur in most arid regions in western North America, and their extensive range extends from southeastern Oregon and southern Idaho down through eastern California, Nevada, western Utah, western Colorado, western New Mexico, western Texas, Arizona, and deep into central Mexico and all through Baja California (Figure 1.4). They are sufficiently abundant in Arizona, Nevada, New Mexico, and Texas that they are legally harvested furbearers. In the states along the margin of the range, kit foxes are less abundant either naturally (e.g., lower habitat suitability along the range margins), anthropogenically (e.g., habitat alteration and loss), or both. Consequently, kit foxes are listed as state endangered in Colorado, state threatened in Oregon, a protected non-game species in Idaho, and a species of concern in Utah. In Mexico, kit fox populations are likely declining due to habitat loss, harvests, and predator control programs, and the species is considered "vulnerable" (List and Cypher 2004). In California, desert kit foxes (subspecies *arsipus*) are relatively widespread and common; however, harvesting or harming desert kit foxes is prohibited, primarily because they cannot reliably be distinguished from San Joaquin kit foxes except through genetic methods.

The taxonomic status of kit foxes in general is still an academic debate (see Evolution and Taxonomy). Regardless, the San Joaquin kit fox is considered a distinct "evolutionary unit," is geographically isolated from other kit foxes, and is considered rare and declining due to continuing habitat loss. Thus, it has been afforded formal protections under both the federal Endangered Species Act and the California Endangered Species Act.

The historical abundance of San Joaquin kit foxes is unknown. An early estimate of 12,134 foxes may have been overly optimistic for a number of reasons (see History of Conservation Efforts). The only current estimate available is one based on remaining high-quality habitat and an assumption of two breeding adults per home range in this habitat. This produces an estimated 1496 breeding adults, but

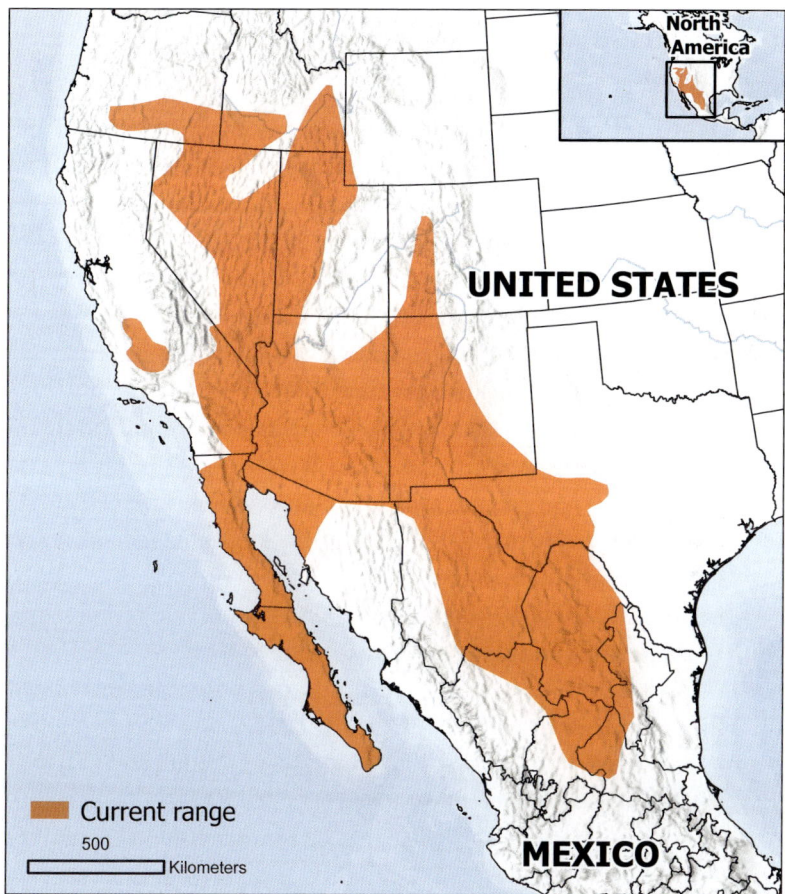

FIGURE 1.4. Range map for kit foxes (modified from List and Cypher 2004). The disjunct part of the range in central California is the range of the San Joaquin kit fox. Prepared by Scott Phillips.

the total number of foxes likely is higher once nondispersing foxes within home ranges, nonresident foxes, young of the year, and foxes residing in urban environments are considered; however, this would still result in a modest number of foxes. The remaining San Joaquin kit foxes persist in three larger "core" populations and fewer than a dozen smaller "satellite" populations with varying levels of connectivity (see Distribution and Habitat Preferences). Genetic exchange still occurs between the populations, although unique alleles do occur in some populations (Schwartz et al. 2005; Wilbert 2013).

EVOLUTION AND TAXONOMY

Kit foxes are in the family Canidae in the order Carnivora. As summarized in Alderton (1994), carnivores were first observed in the fossil record approximately 40 million years before present (mybp). Canids appeared in the late Eocene epoch more than 30 mybp in what is now North America. They resembled mongooses and civets more than modern-day canids. Early canid species began dispersing to other continents toward the end of the Eocene. Canids experienced adaptive radiation in Asia, and their diversity expanded considerably. One of the lineages that evolved eventually gave rise to the two main lineages that became modern-day wolves and foxes. Vulpine foxes first appear in the fossil record in the mid-Miocene epoch approximately 9–12 mybp (Savage and Russell 1983). Animals identified as *Vulpes* were first detected in Pleistocene epoch strata in Eurasia from about 3 mybp. Ancestral canids, including the red fox, *Vulpes vulpes*, then moved across the Bering land bridge back into North America approximately 2 mybp.

Fossil remains of a predecessor to kit and swift foxes have been found in what is now the west Texas/eastern New Mexico region. It is estimated that swift foxes first appeared about 500,000 ybp and that kit foxes first appeared about 100,000–200,000 years ago (Kurtén and Anderson 1980). These foxes then proceeded to spread throughout arid habitats in western North America.

Small arid land foxes in North America were first noted scientifically by the Lewis and Clark expedition in 1806, and from these observations the swift fox was formally described by Thomas Say in 1823 (Dragoo and Wayne 2003). The kit fox was first scientifically described by C. Hart Merriam in the late 1800s (Merriam 1888). He named the fox *Vulpes macrotis*; *Vulpes* is Latin for "fox" and *macrotis* refers to the large ears. Merriam's description was based on a single subadult specimen from southern California. He considered this new fox to be different from the swift fox. Thus, kit foxes and swift foxes were initially recognized as distinct species (Dragoo and Wayne 2003). However, since then, the taxonomy of kit foxes and swift foxes has been moderately dynamic, with revisions at the species and subspecies levels. These revisions are succinctly summarized in Dragoo and Wayne (2003).

As early as 1929, Seton questioned the distinction and suggested that kit foxes might just be a "race" of the swift fox, *Vulpes velox* (Seton 1929). This distinction was further questioned by Hall (1946) and later by Egoscue (1979). Packard and Bowers (1970) reported that the ranges of swift foxes and kit foxes overlap. Rohwer and Kilgore (1973) established that hybridization occurs between the two species in this relatively narrow (ca. 100 km or 60 mi) but apparently stable contact zone in eastern New Mexico and western Texas. They also determined that selection seemed to favor parental forms over hybrids and therefore concluded that recognition of two species was appropriate. This conclusion was further supported by morphological analyses (Packard and Bowers 1970; Thornton and Creel 1975); however, Hall (1981) questioned this conclusion based on the fact that any gene flow was occurring between the two species. Further support for recognition of separate species was provided through additional morphological work (Stromberg and Boyce 1986) and later by mitochondrial DNA analyses (Mercure et al. 1993; Maldonado et al. 1997). However, a case again was made for swift foxes and kit foxes being conspecific based on analyses of morphological data and allozymes (Dragoo et al. 1990) and microsatellite DNA (Dragoo and Wayne 2003). Based on this last analysis, all North American arid lands foxes were considered *Vulpes velox*, and no subspecies were recognized (Dragoo and Wayne 2003).

The San Joaquin kit fox was first noted scientifically at the very beginning of the twentieth century. Naturalists of the period commonly named new species based on variations in pelage, external measurements, and skull measurements, frequently from a small number of specimens or even just one specimen. Thus, Merriam named a new species of fox, *Vulpes muticus*, based on a single specimen from the San Joaquin Valley (Merriam 1902). (The Latin term *muticus* means "docked, curtailed, or cut-off" and probably refers to the diminutive size of the fox.) This large adult male was collected on March 22, 1902, near the town of Tracy in San Joaquin County at the northern end of the Valley and is the type specimen for the San Joaquin kit fox.

Grinnell (1913) compiled a list of California mammals in which he recognized the San Joaquin kit fox as a subspecies of kit fox and named it *Vulpes macrotis muticus*. Ten years later, he updated his list and changed the subspecific epithet from *muticus* to *mutica*, presumably so the gender of the Latin trinomial would be consistent (Grinnell 1923). Based on an analysis of cranial and external morphological measurements, Grinnell et al. (1937) considered all kit foxes in California to be one species, *Vulpes macrotis*, and identified three subspecies: *V. m. mutica*, *V. m. arsipus*, and *V. m. macrotis*. Hall and Kelson (1959) also recognized these same three subspecies along with five others that occurred outside of California. Kit fox subspecies have been redefined on several occasions since (e.g., Waithman and Roest 1977; Mercure et al. 1993), but kit foxes in the San Joaquin Valley typically were always recognized as distinct. However, based on their analysis of morphological and allozyme data, Dragoo et al. (1990) concluded that all kit foxes constituted a single subspecies, *Vulpes velox macrotis*, and San Joaquin kit foxes were not recognized as being distinct from other kit foxes. More recent analyses by Dragoo et al. summarized in Dragoo and Wayne (2003) suggest that no subspecies should be recognized but did acknowledge that San Joaquin kit foxes were the most distinct "genetic population unit."

The debates regarding the taxonomy of kit foxes and swift foxes or of subspecific designations are primarily academic ones. For the purposes of conservation, classification of populations typically trends toward maintaining unique entities as this uniqueness provides justification for conserving these populations. As a prime example, after Dragoo et al. (1990) published their work, which synonymized kit foxes with swift foxes and grouped all kit foxes into just one subspecies, a petition was then submitted in 1992 to the US Fish and Wildlife Service (USFWS) to remove protections for the San Joaquin kit fox under the Endangered Species Act because of lack of taxonomic validity. However, the USFWS concluded that the status of kit fox and swift fox taxonomy remained a subject of ongoing scientific debate and that regardless of the outcome of the continuing debate, the San Joaquin kit fox was considered to constitute a distinct population segment subject to protection under the Endangered Species Act (USFWS 1998). Consequently, the petition was not approved, and *Vulpes macrotis mutica* remains listed as an endangered species.

The California Department of Fish and Wildlife also continues to consider San Joaquin kit foxes a distinct subspecies, and they remain listed as state threatened. Furthermore, the IUCN also continues to recognize San Joaquin kit foxes as a distinct entity, and kit foxes and swift foxes as separate species (Sillero-Zubiri et al. 2004). Finally, the entire concept of species and subspecies has come under greater scrutiny as more advanced systematic techniques emerge, particularly genetic ones, and decisions on where to draw lines of distinction become more difficult (Frankham et al. 2017). However, while taxonomic distinctions become less certain, the concept of Evolutionarily Significant Units, or ESUs (Moritz 1994), is gaining broad acceptance, particularly regarding conservation of populations (Crandall et al. 2000). Regardless of taxonomic classification, by virtue of geographic isolation and at least some degree of genetic differentiation, the San Joaquin kit fox is being treated by conservation agencies as an ESU. Also, because of relative rarity, protections under the Endangered Species Act and California Endangered Species Act remain in effect for San Joaquin kit foxes.

Chapter 2

NATURAL HISTORY

Many aspects of the natural history of San Joaquin kit foxes are similar to those of all kit foxes and in many cases to closely related swift foxes as well. Most of the information presented in this chapter is summarized from publications, reports, and observations on San Joaquin kit foxes. Information on other kit foxes is sometimes used to supplement the San Joaquin kit fox material or to help fill data gaps. Likewise, the information on San Joaquin kit foxes generally can serve as a surrogate for other kit fox populations in situations where data are not currently available for those populations.

DISTRIBUTION AND HABITAT PREFERENCES

Kit foxes are not as widespread as other North American foxes, such as red foxes or gray foxes, primarily because of more limited habitat preferences compared with these other species (Cypher 2003). The range of the San Joaquin kit fox is miniscule relative to the range of other kit foxes and is geographically isolated (see Figure 1.4). San Joaquin kit foxes are found only in the San Joaquin Desert region in central California. How they came to occupy this region is not completely clear. Even within this relatively small region, they do not occur in all habitats. This limited range and narrow habitat preference have resulted in San Joaquin kit foxes being relatively uncommon, even prior to the contemporary habitat loss that has led to their current federal status as Endangered.

Distribution

Kit foxes are found in arid habitats such as those typical of the desert regions of western North America. Along with swift foxes, they are commonly referred to as "arid-land foxes." Also, kit foxes possess adaptations that facilitate their existence in hot, dry environments (see Chapter 1). All of that said, kit foxes exhibit sufficient ecological plasticity that they are capable of living in more mesic environments as well. In addition to arid habitats, kit foxes have occasionally been found in semi-chaparral habitats, mesic grassland habitats, and drier montane habitats near 3000 m (10,000 ft) in the Sierra Nevada. Their range does not appear limited by food resources or feeding strategy. Although kit foxes are foraging specialists on heteromyid rodents (see Foraging Ecology), they also readily exploit a diversity of invertebrates, other rodents, rabbits, birds, reptiles, and other foods. Thus, they can alter their diet based on the availability of foods in a given area. They also do not appear limited by cold temperatures; the range of the species includes regions with cold, sometimes snowy conditions such as southeastern Oregon, southwestern Idaho, northern Nevada, northern Utah, and some high-elevation montane areas (List and Cypher 2004).

Predation risk may be the primary factor limiting and defining the range of kit foxes. Bobcats potentially play a significant role, although this has yet to be confirmed through field studies. If bobcats do limit kit foxes, then this situation may be even a bit more complex in that coyotes may play a role, but not necessarily through their predation on kit foxes. Kit fox mortalities attributable to coyotes appear to be primarily an act of interference competition (see Interspecific Interactions). However, mortalities attributable to bobcats appear to constitute more classical predation since the kit foxes are usually consumed. Coyotes will also kill bobcats (interference competition) as well as compete with them for food, particularly rabbits and rodents (exploitative competition). Thus, coyotes can markedly limit bobcat abundance (Robinson 1961; Litvaitis and Harrison 1989; Anderson and Lovallo 2003). Whereas kit foxes use dens to elude coyotes, bobcats use dense vegetation and steep terrain or climb trees and tall shrubs to elude coyotes. Thus, bobcats are not common in areas with gentle or flat terrain lacking trees or dense shrubs and with sparse ground cover. These are precisely the habitat conditions considered most optimal for kit foxes (Warrick and Cypher 1998; Cypher et al. 2013). Undoubtedly, kit fox foraging efficiency likely increases in these habitat conditions as it is easier to hunt their preferred prey, kangaroo rats; however, reduced bobcat abundance in these areas may be equally important. The relationship between bobcat and coyote interactions and the distribution of kit foxes is a topic that warrants further investigation.

Although kit foxes as a species have a fairly extensive distribution, San Joaquin kit foxes are found only in the San Joaquin Desert region in central California, which is isolated from the rest of the kit fox range. The probable origin of kit foxes in this region is actually one shared by numerous other animal and plant species. Many of the species found in the San Joaquin Desert are closely related to and indeed were derived from Mojave Desert species (Peabody and Savage 1958; Germano et al. 2011). Today, the two deserts are geographically separated by the southern Sierra Nevada and the Transverse Ranges. Although relatively arid, these mountains attain sufficient elevation (>1200 m or 4000 ft) such that they support chaparral, oak woodlands, and conifer forests unsuitable for desert species and therefore form an effective barrier to movement between the two deserts.

The conditions described above that impede movement by desert species across the mountains are geologically somewhat recent. The last glacial maximum (ice age) ended about 15,000 ybp at the end of the Pleistocene epoch, and then a period of glacial retreat began, referred to as the Holocene epoch (Peabody and Savage 1958). Even during the Holocene (which continues to present), climatic conditions varied. Moratto et al. (1978) define three "climatic ages" during the Holocene: the Anathermal (up to ±7500 ybp) when conditions were relatively cool and moist, the Altithermal (±7500–4000 ybp) when conditions were relatively warm and dry, and the Medithermal (±4000 ybp–present) when conditions were much as they are today (absent anthropogenically induced climate change effects). During the Altithermal, conditions over the southern Sierra Nevada and the eastern Transverse Ranges may have been sufficiently desert-like that some Mojave species managed to migrate into the southern San Joaquin Valley. It is also possible that similar migrations may have preceded this. Many of these immigrants persist today. Some examples include allscale saltbush, spinescale saltbush, jackass clover, desert spiny lizard, glossy snake, and Le Conte's thrasher (Germano et al. 2011). There are even still a few clumps of creosote bush in the Elk Hills area in the southwestern San Joaquin Valley (Hinshaw 1997). Fossil remains of desert tortoises have been recovered from the McKittrick tar pits located in the southwestern San Joaquin Valley (Brattstrom 1953). Both creosote bush and desert tortoises are characteristic species of the Mojave Desert.

In the past approximately 4000 years, conditions have not been suitable for continued movement by species (and their genes) between the San Joaquin and Mojave deserts. Thus, with geographic isolation and lack of genetic exchange, many of the Mojave immigrants into the San Joaquin Valley evolved sufficiently to be recognized as new species, subspecies, or varieties, and these forms are ecological equivalents of the original Mojave Desert species (Germano et al. 2011). Some examples of Mojave Desert émigrés and their current-day San Joaquin Desert counterparts include long-nosed leopard lizards and blunt-nosed leopard lizards, white-tailed antelope squirrels and San Joaquin antelope squirrels, beavertail prickly-pear and Bakersfield cactus, desert kangaroo rats and giant kangaroo rats, Merriam's kangaroo rats and San Joaquin kangaroo rats, and of course desert kit foxes and San Joaquin kit foxes.

The San Joaquin Desert is a relatively small desert (Figure 2.1). Most of the desert occurs in the southern and western portions of the San Joaquin Valley, but it also extends to the southwest into the Carrizo Plain and Cuyama Valley. It encompasses approximately 28,493 km^2 in central California. For comparison, the Mojave Desert encompasses 140,000 km^2 (Germano et al. 2011). Consequently, even prior to the habitat destruction that has occurred since Europeans permanently settled in the region, the distribution of the new immigrant species from the Mojave Desert was not extensive. Furthermore, extensive lakes, wetlands, marshes, and riparian areas occurred within this desert, and these habitats were unsuitable for use by the desert-adapted species, further limiting their distribution as well as their overall population sizes (USFWS 1998). Thus, with the profound habitat loss that has occurred, many of the desert-adapted species are now considered rare, with many requiring formal protections. These species include the Bakersfield cactus (federal and state

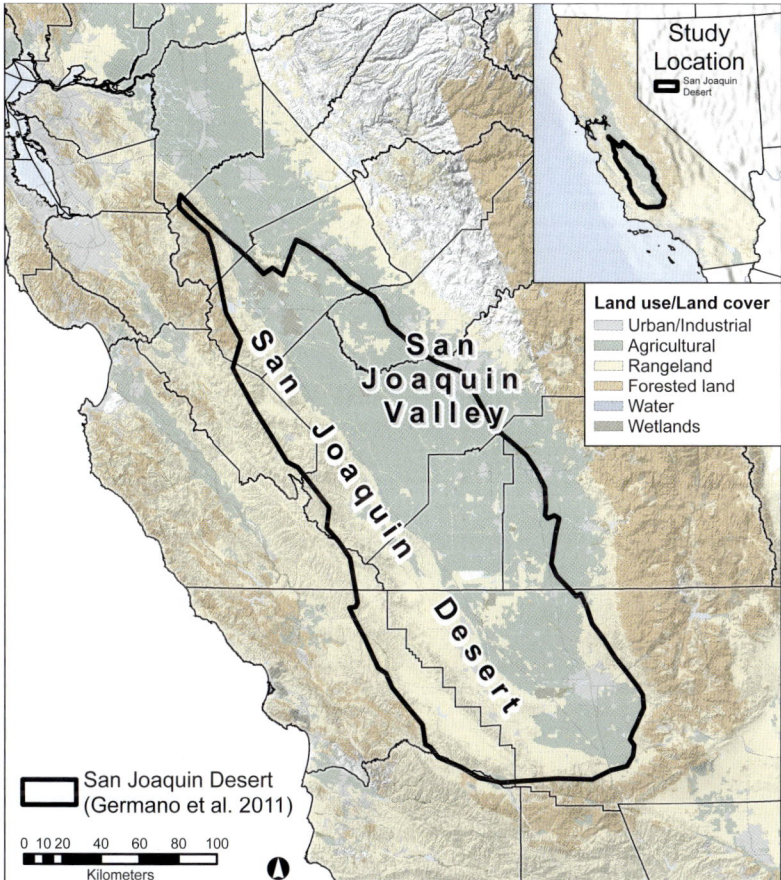

FIGURE 2.1. Location of the San Joaquin Desert in central California. Prepared by Scott Phillips.

Endangered), California jewelflower (federal and state Endangered), San Joaquin woolly-threads (federal Endangered), Kern mallow (federal Endangered), blunt-nosed leopard lizard (federal and state Endangered), San Joaquin antelope squirrel (state Threatened), giant kangaroo rat (federal and state Endangered), Tipton kangaroo rat (federal and state Endangered), Fresno kangaroo rat (federal and state Endangered), and San Joaquin kit fox (federal Endangered and state Threatened).

All the species listed above occur exclusively in the San Joaquin Desert. The San Joaquin Desert is described in detail in Germano et al. (2011). To summarize, the region has a Mediterranean climate characterized by hot, dry summers and cool, wet winters, with frequent fog. Annual average precipitation is typically 230 mm or less, and the average in the very arid southwestern portion of the desert near the towns of McKittrick and Taft is about 117 mm. This precipitation falls almost exclusively as rain during October through April. Soils within this San Joaquin Desert are classified primarily as aridic or xeric. The plant communities are classified as Desert Saltbush Scrub or Desert Sink Scrub (Holland 1986) or Allscale Scrub Alliance (Sawyer et al. 2009). Common shrubs include desert saltbush, spiny saltbush, and alkali goldenbush. A showy display of native annual forbs and grasses is present for a few weeks in the late winter and early spring, but generally the ground cover is dominated by the non-natives red brome, Arabian grass, and red-stemmed filaree.

All the remaining persistent San Joaquin kit fox populations are found within the bounds of the San Joaquin Desert. The largest remaining population occurs in the Carrizo Plain region (Figure 2.2) in eastern San Luis Obispo County (USFWS 1998; Cypher et al. 2013). Roughly estimating, about a third of the remaining kit foxes may occur in this population. Fortunately, much of the best habitat in this region is conserved within the Carrizo Plain National Monument managed by the US Bureau of Land Management (BLM), ecological reserves managed by the California Department of Fish and Wildlife, and conservation lands set aside as mitigation for large solar farms in the northern part of the region (see History of Conservation Efforts). Another almost equally large population of kit foxes occurs in western Kern County (Figure 2.2). The Kern and San Luis Obispo populations are separated by the Temblor Range. This range reaches approximately 1200 m (4000 ft) in elevation. Few if any kit foxes reside in the range; the very rugged upper elevations increase predation risk for kit foxes, as discussed previously. However, kit foxes apparently make it across this range on occasion, as evidenced by foxes detected on cameras near the crest of the range, and also the fact that there appears to be some amount of genetic exchange between the two populations (Wilbert et al. 2020). A considerable proportion of the better habitat in the western Kern County area is conserved in areas managed by a number of different entities; however, much more of the habitat could and should be conserved.

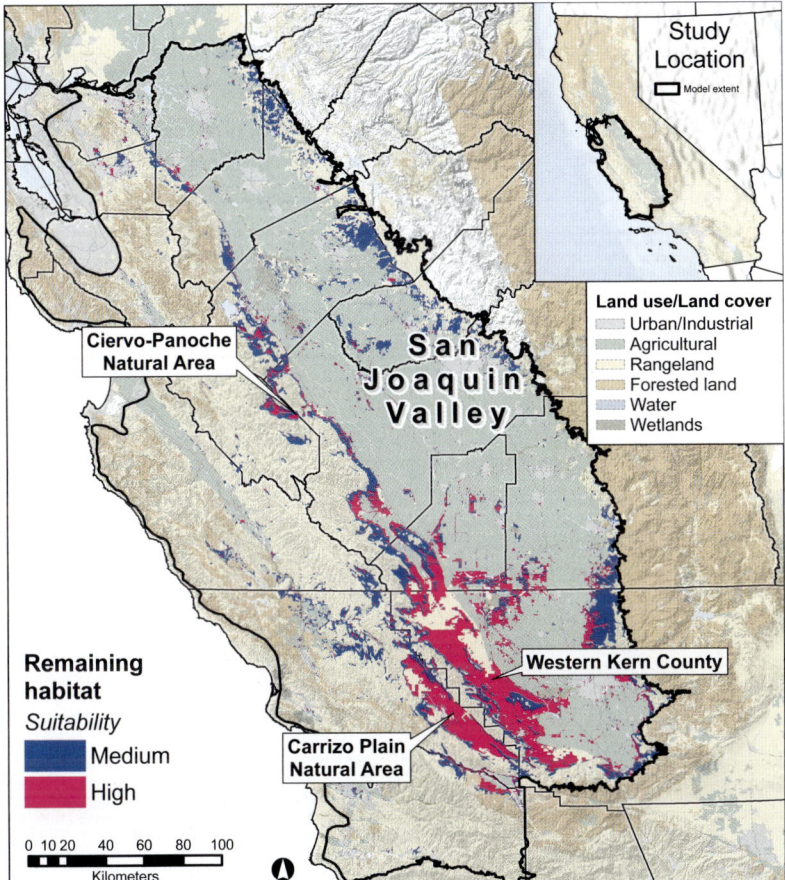

FIGURE 2.2. Locations of core populations of San Joaquin kit foxes. Suitability of remaining habitat is also shown. Prepared by Scott Phillips.

Approximately 160 km (100 mi) to the north of the Carrizo and western Kern County populations, another occurs in the Panoche Valley region (Figure 2.2) in eastern San Benito County (USFWS 1998; Cypher et al. 2013). This population is considerably smaller than the Kern and San Luis Obispo populations and probably numbers about 100–200 individuals. Some of the good habitat in this area is on BLM lands, and just recently, much of the most highly suitable habitat in the region was conserved and is being managed by the Center for Natural Lands Management. Being the largest remaining populations, at least among those in natural lands, the Kern, Carrizo, and Panoche Valley populations were designated "core" populations in the recovery plan for kit foxes (USFWS 1998) and are considered essential to the conservation and recovery of San Joaquin kit foxes.

In addition to these three core populations, there are several satellite populations (Figure 2.3) scattered around the range (USFWS 1998; 2010), each with probably around a dozen to 100 adult foxes. At the north end of the range, one of these populations occurs primarily on private lands in western Merced County. These lands are located south of State Route 152 and west of Interstate 5, and only a small proportion of the lands currently (as of August 2023) have any sort of conservation status. This population is the most northern extant population. It is located approximately 65 km (40 mi) north of the Panoche population. There is some connectivity between the two areas, and some genetic and demographic exchange likely occurs between the two. Approximately 65 km (40 mi) southeast of the Panoche Valley, a population of kit foxes occurs on lands around the city of Coalinga. These lands are largely active or abandoned oil and gas production lands that are owned by various oil companies. As with the western Merced area, only a small proportion of the lands currently (as of August 2023) have any sort of conservation status. A small number of foxes also occur in the Kettleman Hills just to the south. Another population is located in the Semitropic region in northern Kern County, east of Interstate 5 and from State Route 46 north to the area around Kern National Wildlife Refuge. A considerable proportion of the best habitat in this area is conserved and managed by several different conservation organizations. This area potentially has some connectivity with the north end of the western Kern core population, although Interstate 5 constitutes a considerable barrier between the two.

A fourth satellite population occurs in the Cholame Valley area, located in the very southeast corner of Monterey County where it abuts the very northeast corner of San Luis Obispo County. This area may have some connectivity with the Carrizo area, about 32 km (25 mi) to the south, and the western Kern area, about 32 km to the southeast; however, busy roads and inhospitable topography present challenges to foxes moving between this satellite population and either of the two core areas. Considerable suitable kit fox habitat has been conserved in this area in the past few years as mitigation for solar farms and other projects. Farther south, a small satellite

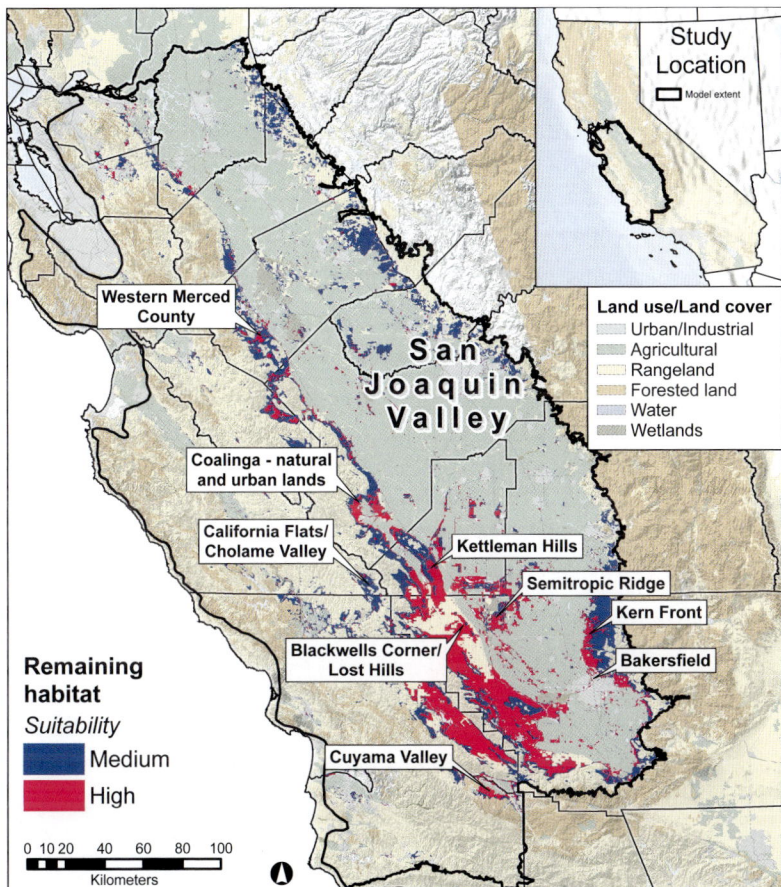

FIGURE 2.3. Locations of satellite populations of San Joaquin kit foxes. Suitability of remaining habitat is also shown. Prepared by Scott Phillips.

population currently (as of August 2023) persists in the Cuyama Valley, located on the boundary between southeastern San Luis Obispo County and northeastern Santa Barbara County. Physically, this population is only about 4 km (2.5 mi) away from the Carrizo core area; however, the rugged and inhospitable Caliente Range, with elevations over 1500 m (5000 ft), separates the Cuyama Valley from the Carrizo Plain. Kit foxes may occasionally move between the two areas by traveling around the southern end of the Caliente Range. Other than a few scattered small BLM parcels, very little of the remaining habitat in the Cuyama Valley is conserved.

In the center of the southern San Joaquin Valley in Kern County, a number of kit foxes occur on lands between the city of Bakersfield and Interstate 5. Much of the habitat in this area is conserved. This area is essentially contiguous with the western Kern core area; however, Interstate 5 and developments along this highway corridor are increasingly reducing connectivity between this area and the core area. On the east side of the San Joaquin Valley, there is a satellite population in the Kern Front area that lies between the city of Bakersfield and Poso Creek. Most of this area is within an active oil and gas production area, and little if any of the habitat is conserved.

Small pockets of foxes are known to occur in other locations. These pockets do not constitute persistent populations, but instead are areas that foxes, sometimes just a single family group, colonize and occupy for some period of time, sometimes even less than a year. The foxes then disappear from the area, and it may be some time before it is occupied again. One such area is located in hilly terrain just east of the city of Bakersfield. Another is located north of Poso Creek in a narrow band of grassland between State Route 65 and the foothills of the Sierra Nevada. This area extends from Poso Creek about 24 km (15 mi) up to the boundary between Kern and Tulare counties.

Interestingly, the largest satellite population is the urban kit fox population inhabiting the city of Bakersfield (see Kit Foxes and Urban Areas). This population may even be larger than the Panoche core population. It is demographically quite robust (Cypher 2010). Also, the city of Bakersfield is growing rapidly, and kit foxes quickly colonize the newly created urban environments. Thus, this population is the only one of all the San Joaquin kit fox populations that is actually increasing in size.

San Joaquin kit foxes have also been documented in other locations, sometimes in sizeable numbers. A particularly interesting example is the Camp Roberts area in the Salinas Valley on the border between San Luis Obispo County and Monterey County (USFWS 1998, 2010). In the 1980s and 1990s, this population was quite large and at times exceeded 100 individuals (Berry and Standley 1992; White et al. 2000). Demographically, the population seemed quite robust with good survival (Standley et al. 1992) and reproductive rates (Spencer et al. 1992); however, in the late 1990s, this population began to decline, and by the very early 2000s, no individuals were detected during annual monitoring at Camp Roberts (an Army

National Guard military base) or other locations (e.g., Fort Hunter-Liggett) in the Salinas Valley where foxes had previously been observed. A report by Balestreri (1981) implies that kit foxes were quite uncommon in the Salinas Valley region prior to the 1960s. Indeed, Grinnell et al. (1937) make no mention of kit foxes being present in this region. However, kit foxes were more commonly documented from the 1960s on, and as mentioned above, the population at Camp Roberts grew quite large. One potential explanation is that environmental changes in the Salinas Valley, particularly in the mid-twentieth century, may have facilitated colonization and occupation of this region by kit foxes. These changes may have included over-grazing of rangelands (Jensen 1972) and frequent wildfires (both of which would have reduced vegetation density) and more aggressive coyote control, particularly by USDA predator control programs (Balestreri 1981). These more open habitat conditions and fewer large predators would have made the region more suitable for kit foxes. However, in the latter part of the century, grazing practices were better managed (or ceased entirely on some lands), and lethal predator control programs were significantly scaled back or terminated; these factors would have resulted in denser vegetation conditions and a greater abundance of larger predators. Also, more and more habitat was being converted to agriculture, particularly vineyards, and residential areas. Thus, available habitat was rapidly disappearing, with the remainder increasingly fragmented.

Another factor that may have negatively affected kit foxes in the Salinas Valley was the appearance and increased abundance of non-native red foxes (Lewis et al. 1999). Normally, red fox abundance is suppressed due to predation by coyotes (Lewis et al. 1999; Cypher, Clark, et al. 2001); however, red foxes can persist and thrive in agricultural and residential areas where coyote abundance is typically much reduced (Lewis et al. 1999; Clark et al. 2005; Cypher 2010). Red foxes have been documented killing kit foxes (Ralls and White 1995; Clark et al. 2005) and likely compete with kit foxes for food and dens. Finally, in the late 1990s, an epizootic of rabies occurred in the striped skunk population in the Salinas Valley region, and this may have impacted other species such as kit foxes (White et al. 2000). Indeed, rabies was determined to be the cause of death of at least two kit foxes at Camp Roberts in 1990 (Standley et al. 1992). Kit foxes may have been particularly vulnerable, as skunks occasionally use kit fox dens, sometimes even ones currently occupied by kit foxes (Harrison et al. 2011). Thus, the potential for transmission of disease is high. However, if the kit fox population was significantly impacted by rabies, it may have simply hastened what was an inevitable outcome, given the downward trend in kit fox numbers in the Salinas Valley in the 1990s.

Another area of interest is one referred to as the "northern range". This is the region from northwestern Merced County (north of State Route 152) north, including Stanislaus, San Joaquin, Contra Costa, Alameda, and Santa Clara counties. The type locality for the San Joaquin kit fox, near the city of Tracy in San Joaquin County, is within this region; however, this was based on the collection of one specimen, and no other information is provided on the abundance of kit foxes in this area. In 1902, it is likely that considerably more potential habitat was available in this region, thereby increasing the possibility that kit foxes might occur in or at least visit this area. However, Grinnell et al. (1937) did not detect kit foxes in this region and reported that kit foxes were likely extinct in that area. There is no historical evidence that large or viable kit fox populations ever occurred in this area. The region was surveyed extensively with trained scat detection dogs during 2001–2003, but no kit fox scats were found, although red fox scats were common (Smith et al. 2006).

Kit foxes are occasionally observed in this northern region (Constable et al. 2009), and occasional reproduction has even been documented (Hall 1983; Orloff et al. 1986). However, habitat conditions in this region are generally suboptimal for kit foxes (Cypher et al. 2013). In particular, the vegetation in this region is dominated by nonnative grasses, particularly wild oats, that form dense ground cover that is also relatively tall compared with the size of a kit fox. Thus, kangaroo rats (preferred prey for kit foxes) are not abundant, and kit foxes are more easily ambushed by predators. The majority of kit foxes detected in this region are likely individuals dispersing from southern populations that unfortunately ended up in an unsuitable area. In that sense, the northern range may actually function as a population sink for kit foxes. In addition, there is some evidence that a number of the kit fox occurrences in the northern range could actually be misidentified coyote pups or other fox species (Sproul and Flett 1993; Clark et al. 2007).

San Joaquin kit foxes occasionally show up in other nontypical locations as well, including the vernal pool grasslands in the northeastern part of the San Joaquin Valley, the Shell Creek area in the Coast Ranges in San Luis Obispo County, and the Hollister area in San Benito County. However, there is no evidence that these observations are of resident or reproducing foxes; it is more likely they are dispersing individuals. Such observations seem more common when fox numbers in the core and satellite populations are elevated, "surplus" foxes are present, and the habitat is saturated with resident foxes. Caution must be exercised that observations of foxes in unusual areas are not construed as indicating that those areas contain suitable habitat to support persistent fox populations. This could divert conservation efforts from areas that truly constitute high-quality habitat and where habitat protection is more likely to significantly contribute to the conservation and recovery of San Joaquin kit foxes.

Almost all the remaining suitable habitat and all the known persistent fox populations occur within the area delineated as the San Joaquin Desert, described previously. The boundaries of the range of the San Joaquin kit fox are generally little changed from what they were before the alteration of habitats by European and other settlers. However, within those boundaries there has been a profound loss or alteration of suitable habitat, which has led to the San Joaquin kit fox, and many co-occurring species, being imperiled.

Habitat Suitability

In the previous pages, suitable habitat for the San Joaquin kit fox is extensively described, along with some hints as to what constitutes suitable habitat. Initial research on San Joaquin kit fox habitat attempted to characterize it based on general habitat descriptions or by the plant community associations in which foxes were observed. For example, Grinnell et al. (1937) reported that the San Joaquin kit fox was found in "the dry plains of the San Joaquin Valley." Laughrin (1970) described suitable habitat generally as arid grasslands, sometimes with shrubs such as desert saltbush, and the alkali sink community. In the species account in the recovery plan that includes kit foxes, the authors even attempted to list the ecological communities where kit foxes were found, based primarily on the assumed range for San Joaquin kit foxes (USFWS 1998, 129). Thus, in the southern portion of the range, the kit fox was said to be commonly associated with Valley Sink Scrub, Valley Saltbush Scrub, Upper Sonoran Subshrub Scrub, and Annual Grassland. In the central portion of the range, the kit fox was said to be associated with Valley Sink Scrub, interior Coast Range Saltbush Scrub, Upper Sonoran Subshrub Scrub, Annual Grassland, and the remaining native grasslands. In the northern portion of the range, the kit fox was said to be associated with annual grassland and Valley Oak Woodland. Finally, the authors listed "other" plant communities in the San Joaquin Valley providing kit fox habitat, including Northern Hardpan Vernal Pool, Northern Claypan Vernal Pool, Alkali Meadow, and Alkali Playa.

The long list of communities above encompasses all the natural habitats where kit foxes have been observed. However, kit foxes do exhibit ecological plasticity and can also travel considerable distances, particularly when dispersing (see Space Use and Movements). Therefore, they are occasionally found in areas that may not constitute optimal habitat. For example, no kit fox populations are known to occur in the "other" communities mentioned in the recovery plan (e.g., Northern Hardpan Vernal Pool, Northern Claypan Vernal Pool, Alkali Meadow, and Alkali Playa) despite that fact that considerable amounts of these habitats are still present in the northeastern portion of the San Joaquin Valley. Also, there is no evidence that populations historically occurred in these areas, although, to reiterate, occasional individuals indeed have been documented in these habitats.

Given the profound loss of natural habitats in the San Joaquin Valley and also the limited resources available to conserve them, Cypher et al. (2013) undertook an effort to more quantitatively define the most suitable habitat for San Joaquin kit foxes in the hope that conservation efforts could target this habitat for protection. Habitat suitability for a given species is commonly based on analyzing locations of individuals with radio transmitters and comparing the habitat attributes of areas used by the individuals with areas that are not used. This approach has some inherent biases and limitations, and it also requires a robust quantity of location data from representative areas throughout the range. Such data were not available for San Joaquin kit foxes, so the researchers took a different approach. They identified areas where San Joaquin kit fox populations were present and known to be persistent. That is, areas where kit foxes had been and could be reliably found, and where consistent reproduction was known to occur. These areas were assumed to have highly suitable habitat for kit foxes. Examples of sites with persistent kit fox populations included the valley floor and Elkhorn Plain areas of the Carrizo Plain National Monument in eastern San Luis Obispo County, the Buena Vista Valley and Lokern Natural Area in western Kern County, and the Panoche Valley in southeastern San Benito County. They also identified areas where kit foxes were occasionally reported, but where occurrence did not appear persistent and reproduction was never or at least rarely documented. These areas were considered of moderate suitability for kit foxes. Examples of sites with medium habitat suitability included the Allensworth Ecological Reserve in Tulare County, Camp Roberts in San Luis Obispo County, and grasslands near the San Luis Reservoir in Merced County. They then measured various habitat attributes in these areas to identify suitable conditions for kit foxes.

The three main attributes measured were land use or cover type, terrain ruggedness, and vegetation density. These attributes had been previously identified as important for kit foxes (Grinnell et al. 1937; White et al. 1995; USFWS 1998; Warrick and Cypher 1998; Cypher et al. 2000; Smith et al. 2005; Warrick et al. 2007). Agricultural lands and urbanized areas (but see Kit Foxes and Urban Areas) were excluded from the study. Areas with persistent kit fox populations usually have desert or spiny saltbush scrublands and grasslands dominated by red brome, whereas areas with occasional kit fox presence include alkali sink scrublands and grasslands dominated by wild oats. Areas with persistent populations are also generally characterized by flat or gently rolling terrain (average slopes <5%), and suitability declines as terrain ruggedness and average slope increase, apparently due to an associated increase in predation risk for kit foxes (Warrick and Cypher 1998). Finally, kit foxes are optimally adapted to arid environments with sparse vegetation and a high pro-

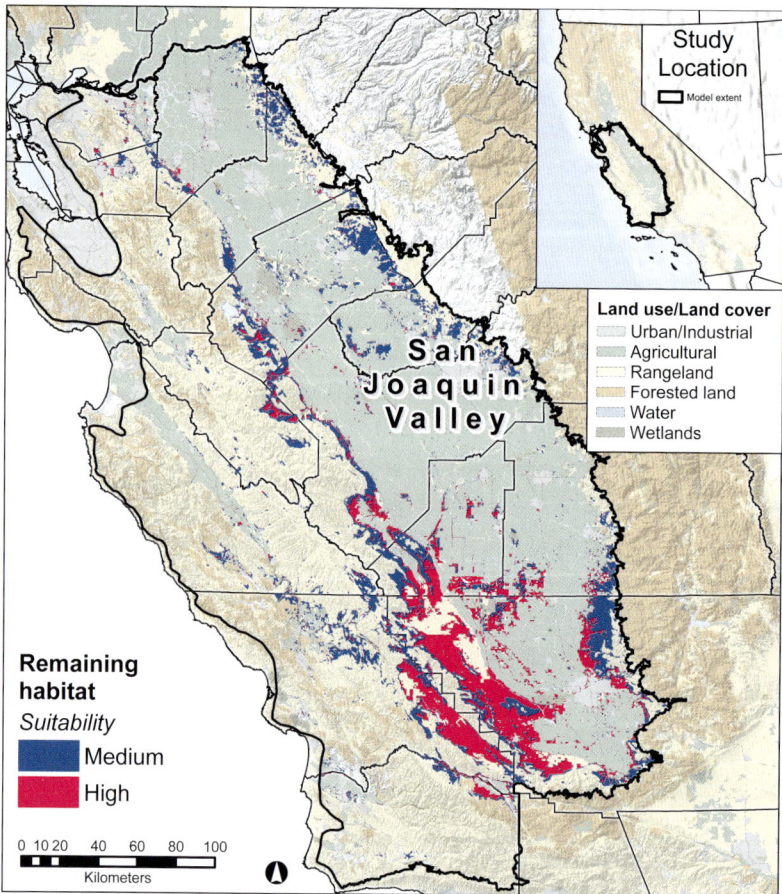

FIGURE 2.4. Remaining suitable habitat for San Joaquin kit foxes in central California. Prepared by Scott Phillips.

portion of bare ground (Grinnell et al. 1937; McGrew 1979). Thus, habitat suitability decreases as vegetation density increases.

These attributes were included in a spatially explicit GIS model which resulted in a map of habitat suitability for San Joaquin kit foxes (Figure 2.4; Cypher et al. 2013). The implications of the amount and distribution of remaining suitable habitat are discussed in History of Conservation Efforts. The purpose here is simply to define the characteristics of suitable habitat for kit foxes. To be succinct, the most suitable habitat conditions are found in areas of flat or very gently rolling terrain with relatively sparse ground cover and considerable bare ground (Figure 2.5). The vegetation communities in the

FIGURE 2.5. Highly suitable habitat for San Joaquin kit foxes: (a) Carrizo Plain in eastern San Luis Obispo County and (b) Lokern area in western Kern County. Photos by Brian Cypher.

suitable areas are characterized as arid scrublands or grasslands. As terrain becomes more rugged and vegetation density increases, the suitability of the habitat for kit foxes declines (Figure 2.6).

The conditions most suitable for kit foxes are essentially those found in arid regions. Consequently, all highly suitable habitat for

FIGURE 2.6. Low suitability habitat for San Joaquin kit foxes. Photo by Brian Cypher.

San Joaquin kit foxes is encompassed within the San Joaquin Desert (Germano et al. 2011). This region generally receives less than 23 cm (9 in) of precipitation annually, naturally limiting ground cover density. The low sparse ground cover facilitates the detection of approaching predators. One of the other reasons these conditions contribute to high suitability for kit foxes is that these same conditions constitute highly suitable habitat for kangaroo rats, which make up the primary prey for kit foxes (see Foraging Ecology). Kit foxes are most abundant in areas where kangaroo rats are also abundant (Grinnell et al. 1937; Cypher et al. 2013).

Common and easily recognized ground cover plants found in highly suitable kit fox habitat include red brome, Arabian grass, red-stemmed filaree, fiddlenecks, tarweeds, goldfields, and phacelias. An interesting dynamic exists with red brome. This is a non-native grass introduced into California with livestock from Europe beginning in the 1600s (Baker 1978; Minnich 2008). It is invasive and found the San Joaquin Valley and other large regions of California to its liking. An ecological consequence is that ground cover is likely much denser in much of the San Joaquin Valley compared with pre-invasion conditions. On one hand, kangaroo rats readily consume the seeds of this grass, and its presence may even be responsible for more consistent food availability and higher kangaroo rat abundance in some areas. On the other, in years of abundant precipitation, red brome forms dense continuous cover that not only excludes native plants, but also is detrimental to kangaroo rats (Germano et al. 2001). These animals, adapted to sparse vegetation conditions with abundant bare ground, have a difficult time moving through this vegetation, enhancing their risk of predation (Williams and Germano 1992; Germano et al. 2001). Also, during the winter, this dense cover can hold abundant moisture, causing the animals to become soaked. During a wet, cold winter in 1994–1995, biologists reported finding kangaroo rats with some sort of respiratory condition, and kangaroo rat abundance declined sharply after that winter (Single et al. 1996). Kit fox abundance declined in response (Cypher et al. 2000). There is a cumulative impact associated with dense red brome in that the dead stems are persistent and form a dense thatch that further inhibits native plants, impedes kangaroo rats, and promotes the growth of red brome the next year. Persistent high density of non-native grasses in some areas appears to have sufficiently reduced habitat quality such that use of these areas by kit foxes has declined or ceased entirely. At one time, kit foxes were commonly observed at the Pixley National Wildlife Refuge and the Allensworth Ecological Reserve in Tulare County; however, during a multi-year period of high precipitation in the early 1990s, these areas became densely covered with red brome and kit fox observations rapidly declined. Currently, kit foxes are rarely observed in these areas and those occasionally detected are likely dispersing animals that do not persist.

Related areas in the San Joaquin Valley receiving more than about 23 cm (9 in) of rain annually usually have dense cover of wild oats. This non-native grass is taller than red brome and can be equally dense. Kit foxes (and kangaroo rats) are only rarely observed in areas dominated by wild oats (Cypher et al. 2013). An interesting consideration is whether these areas may have been suitable for kit foxes (and kangaroo rats) prior to invasion by wild oats.

Shrubs are not a necessary component of kit fox habitat. Kit foxes are extremely abundant in some areas with few or no shrubs. Kangaroo rats do not require shrubs, and neither do kit foxes. In fact, if shrubs are too dense, they actually decrease habitat suitability for both kangaroo rats and kit foxes. Giant kangaroo rats and San Joaquin kangaroo rats in particular become less abundant as shrub density increases (Grinnell 1932; Williams 1992; USFWS 1998). In addition, shrubs provide cover for kit fox predators such as bobcats and coyotes (Warrick and Cypher 1998). These larger predators do not use burrows outside the young-rearing season, and they typically rely on shrubs for shade from the abundant sunshine and warm temperatures typical of the San Joaquin kit fox range. Rabbits, a primary prey of coyotes and bobcats, are also more abundant in areas with higher shrub densities (Nelson et al. 2007; Cypher et al. 2018). One estimate is that kangaroo rat and kit fox abundance tend to begin decreasing as shrub cover rises above about 23% (G. Warrick, CNLM, personal communication). Because of the arid habitat conditions, the most common shrubs in highly suitable San Joaquin kit fox habitat tend to be desert saltbush, spiny saltbush, jointfir, and alkali goldenbush.

One final note: kit foxes do not appear to be directly associated with particular soil types. Generally, they are more abundant in areas with loose-textured soils (Grinnell et al. 1937; Hall 1946; Egoscue 1962; Morrell 1972) but are found on virtually every soil type. Dens appear to be scarce in areas with shallow soils because of the proximity to bedrock (O'Farrell and Gilbertson 1979; O'Farrell et al. 1980), high water tables (McCue et al. 1981), or impenetrable hardpan layers (Morrell 1972). However, kit foxes appear able to establish dens in a variety of soil types, so den excavation is not a limiting factor. Areas considered highly suitable for kit foxes generally have sandy loam soils, primarily because such soils are most suitable for kangaroo rats. Kangaroo rats can easily burrow in these soils, which are typically found in more upland terrain that does not flood and where the soils do not become super-saturated, which decreases thermal insulation and increases potential for any seed stores to spoil (USFWS 1998).

SURVIVAL AND MORTALITY FACTORS

Survival is one of the primary demographic parameters (along with reproduction) that drive the population dynamics of species. Survival and causes of mortality for San Joaquin kit foxes vary spatially with the ecological conditions (e.g., habitat attributes, food availability, predator abundance) present in different areas throughout the range, and survival also varies temporally as ecological conditions vary over time, primarily in response to annual precipitation trends and the concomitant effects on prey availability. However, in examining data available from various studies, some general trends become obvious.

Survival Rates

Numerous studies have assessed survival among San Joaquin kit foxes at several locations throughout the range. Summarizing this information is a bit challenging because of the diversity of methods used to assess survival. Mean annual survival rates vary considerably (Table 1). Reported mean rates for adult foxes (typically ≥1 year old; probability of surviving 365 days) range from 0.05 to 1.0. This variation in rates is substantial and reflects the marked spatial and temporal fluctuations in environmental conditions, particularly food and competitor availability. In general, survival rates are lower when food availability is low. This is likely due to foxes having to spend more time foraging and traveling greater distances to find food. Spending more time out of their dens and traveling farther increase the potential for encountering predators. As further evidence of this, home range sizes are larger in years and in areas with low prey availability (see Space Use, Movements, Dispersal, and Activity). Another general observation is that survival is typically higher in kit fox population core areas compared with satellite population areas. The highest survival rates were observed in the Panoche Valley, Carrizo Plain, and western Kern County core areas, while some of the lowest rates were observed in the Fort Hunter Liggett and western Merced County satellite population areas. Habitat quality, particularly food availability, tends to be higher in core areas while predation risk tends to be higher in satellite population areas, usually

TABLE 1. *Survival rates for adult San Joaquin kit foxes from locations throughout the range*

Location	Year(s)	No. foxes	No. deaths	Mean annual rate	Range	Source	Notes
Fort Hunter Liggett, southwestern Monterey Co.	1990–91	11	8	0.05[a]	–	JSA 1995	Satellite population area; natural lands with occasional military training activities; probability of surviving 274 d
Fort Hunter Liggett, southwestern Monterey Co.	1986–87	4	2	0.28[a]	–	O'Farrell et al. 1987	Satellite population area; natural lands with occasional military training activities; probability of surviving 173 d
Tupman-Buttonwillow, western Kern Co.	1977	13	6	0.33	–	Knapp 1978	Core population area but lands actively being converted to agriculture; probability of surviving 227 d
Elk Hills, western Kern Co.	1980–86	154	67	0.38	0.19–0.70	Cypher et al. 2000	Core population area; natural lands
Western Merced Co.	1985–87	14	7	0.40[b]	–	Briden et al. 1992	Satellite population area; natural lands
Semitropic Ridge, northern Kern Co.	2011–12	13	4	0.45[a]	–	Cypher, Westall et al. 2014	Satellite population area; natural lands
Topaz Solar Farms, San Luis Obispo Co.	2014–17	35	16	0.49	0.32–0.78	Cypher, Westall et al. 2019	Core population area; natural lands; reference site for Topaz Solar Farm study
Lokern Natural Area, western Kern Co.	1989–93	54[c]	31 (all ages)	0.52	0.29–0.76	Spiegel 1996	Core population area; natural lands
Camp Roberts, northern San Luis Obispo Co.	1988–91	67	35	0.53[b]		Standley et al. 1992	Satellite population area; natural lands with occasional military training activities
Elk Hills, western Kern Co.	1980–86	127	47	0.57	0.32–0.86	Cypher et al. 2000	Core population area; oil and gas production
Lokern Natural Area, western Kern Co.	1989–93	49[c]	29 (all ages)	0.58	0.38–0.77	Spiegel 1996	Core population area; oil and gas production

Location	Year(s)	No. foxes	No. deaths	Mean annual rate	Range	Source	Notes
Carrizo Plain, eastern San Luis Obispo Co.	1989–91	33	22 (all ages)	0.61[b]	–	Ralls and White 1995	Core population area; natural lands
Northern Carrizo Plain, San Luis Obispo Co.	2012–13	10	2	0.64[a]	–	Cypher, Fiehler et al. 2014	Core population area; natural lands
Topaz Solar Farms, San Luis Obispo Co.	2014–17	17	7	0.65	0.54–0.85	Cypher, Westall et al. 2019	Core population area; solar site
California Valley Solar Ranch, San Luis Obispo Co.	2014–17	26	9	0.66	0.63–0.68	HTH 2019	Core population area; natural lands; reference site for California Valley Solar Ranch
Carrizo Plain, San Luis Obispo Co.	2015–16	9	2	0.66[a]	–	Cypher, Westall et al. 2022	Core population area; natural lands
Panoche Valley, eastern San Benito Co.	2019–22	23	10	0.66	0.50–0.79	Cypher, Westall, et al. 2023	Core population area; solar site

Notes: The rate is the probability of surviving for 365 days unless otherwise indicated. To facilitate comparison, rates were calculated using the method of Heisey and Fuller (1985), except for the rates reported for the Lokern Natural Area by Spiegel (1996) in which the Kaplan–Meier method (White and Garrott 1990) was used. Data are ordered from lowest to highest Mean annual rate.

[a]One-year study.
[b]Survival rate is for all years combined.
[c]Sample included juveniles (<1 year old) as well as adults.

TABLE 2. *Survival rates for juvenile San Joaquin kit foxes from locations throughout the range*

Location	Year(s)	No. foxes	No. deaths	Mean annual rate	Range	Source	Notes
Western Merced Co.	1985–87	14	10	0.09	–	Briden et al. 1992	Satellite population area; natural lands; estimate is for entire study
Elk Hills, western Kern Co.	1980–86	127	51	0.14	<0.01–0.27	Cypher et al. 2000	Core population area; oil and gas production; 1 May–15 Feb
Elk Hills, western Kern Co.	1980–86	43	35	0.2	<0.01–0.43	Cypher et al. 2000	Core population area; natural lands; 1 May–15 Feb
Camp Roberts, northern San Luis Obispo Co.	1988–91	27	14	0.25	–[a]	Standley et al. 1992	Satellite population area; natural lands with occasional military training activities
Carrizo Plain, eastern San Luis Obispo Co.	1989–91	17	–[a]	0.41	–[a]	Ralls and White 1995	Core population area; natural lands; low prey availability due to drought
Lokern Natural Area, western Kern Co.	1989–93	–[a]	–[a]	0.41	0.15–0.75	Spiegel 1996	Core population area; oil and gas production; Aug–Dec survival
Lokern Natural Area, western Kern Co.	2001–04	35	7	0.48	0.22–0.64	Nelson et al. 2007; Cypher et al. 2009	Core population area; natural lands
Bakersfield, Kern Co.	1997–2003	135	37	0.53	0.30–0.71	Cypher, unpubl. data	Satellite population area; urban
Lokern Natural Area, western Kern Co.	1989–93	–[a]	–[a]	0.57	0.50–0.67	Spiegel 1996	Core population area; natural lands; Aug–Dec survival

Notes: The rate is the probability of surviving for 365 days unless otherwise indicated. To facilitate comparison, rates were calculated using the method of Heisey and Fuller (1985), except for the rates reported for the Lokern Natural Area by Spiegel (1996) in which the Kaplan–Meier method (White and Garrott 1990) was used. Data are ordered from lowest to highest Mean annual rate.
[a] Data not provided in original source.

due to denser vegetative cover (see Distribution and Habitat Preferences). Interestingly, survival tends to be higher in modified habitats. For example, survival rates among urban foxes in Bakersfield were among the highest reported. Also, rates for foxes using solar farms and oilfields were typically higher than rates in nearby natural lands, possibly due to a lower abundance of kit fox predators, particularly coyotes and bobcats, in modified habitats as a result of greater human activity. Other factors clearly influence kit fox survival as well, such as random effects, as indicated by the variation in the rates from Bakersfield (0.55–1.0), where food availability is consistently high (Cypher 2010).

Fewer data are available for juvenile kit foxes, but unsurprisingly, annual survival rates are lower compared with those for adults (Table 2). Mean annual rates varied from 0.09 to 0.57. Juvenile foxes are less experienced at eluding predators than adults. Also, many juveniles are killed during dispersal when traveling through unfamiliar areas where they do not know the locations of dens, which are their primary escape cover.

Mortality Sources

In non-urban habitats, sources of mortality for San Joaquin kit foxes include larger predators, vehicles, disease, toxins, shooting, and accidents. In these non-urban areas, larger predators are typically the most common cause of kit fox mortality. The main predators are coyotes and bobcats. When coyotes kill kit foxes, it appears to be a form of interference competition by an intraguild competitor, because the kit fox carcass is often not consumed (Figure 2.7). The fox is killed, and the carcass is then abandoned. Coyotes exhibit this same form of competition with swift foxes, red foxes, and gray foxes (Cypher 2003). Clearly, kit foxes are able to coexist with coyotes because coyotes are present and abundant in all natural habitats where kit foxes occur. Indeed, kit fox abundance did not increase in response to a coyote control program conducted at the Naval Petroleum Reserves (Cypher and Scrivner 1992). When bobcats kill kit foxes, it appears to represent more classic predation, because the carcass is typically at least partially eaten and the remains commonly cached to be consumed at a later time (Figure 2.8). Bobcats are usually more

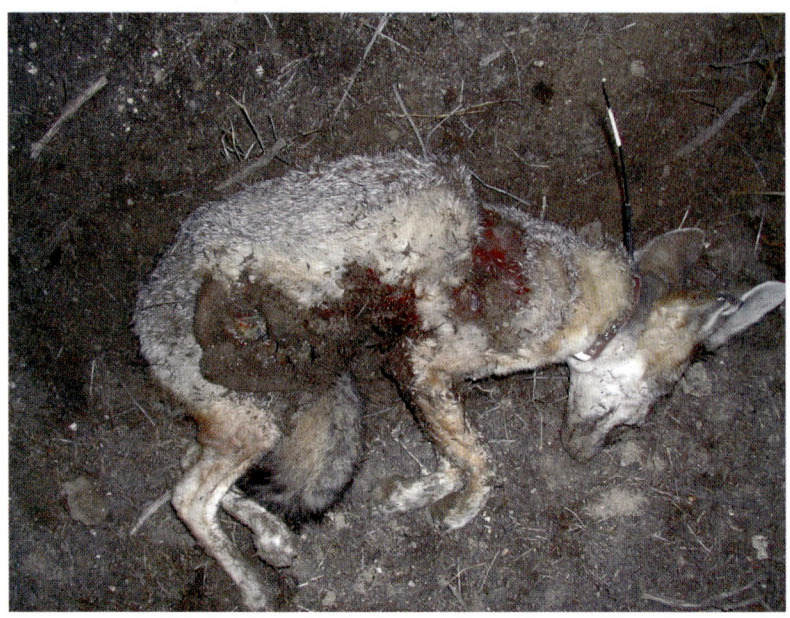

FIGURE 2.7. San Joaquin kit fox killed by a coyote. Photo by Brian Cypher.

FIGURE 2.8. San Joaquin kit fox killed by a bobcat. Photo by Brian Cypher.

FIGURE 2.9. San Joaquin kit fox killed by a golden eagle. Photo by CSUS ESRP.

abundant in areas with denser cover or more rugged terrain. In such areas, kit fox mortality from bobcats can be substantial. Bobcats were the primary source of kit fox mortality at study sites in western Kern County (Spiegel 1996) and eastern San Luis Obispo County (Cypher, Westall, et al. 2019) that included rugged terrain. As mentioned (see Distribution and Habitat Preferences), bobcats might significantly influence the distribution of kit foxes.

Golden eagles also occasionally kill kit foxes (Cypher, Spencer, et al. 2019). Again, this constitutes predation as the carcass is commonly fed upon (Figure 2.9). This predation appears to have a seasonality to it, with eagles taking foxes mostly in the spring. The reason for this is twofold. Although golden eagles can kill adult foxes, they seem to go primarily after pups, which are smaller and easier to kill. The pups begin emerging from natal dens in March and by midsummer are becoming sufficiently large that they may not be easy to capture and kill. Most of the documented predation on San Joaquin kit foxes occurred in April and June (Cypher, Spencer, et al. 2019). The second reason for the seasonality is that late winter and spring are when golden eagles are raising chicks and therefore have higher food demands. Thus, they may be foraging more intensely.

Red foxes are another larger intraguild competitor with kit foxes. Red foxes are not native to the range of San Joaquin kit foxes, but introduced red foxes have spread throughout much of California, including the San Joaquin Valley. As a result of interference competition from coyotes, red foxes are found primarily in agricultural areas and areas with human activity, where coyotes are less abundant. Red foxes are uncommon or even absent in most of the natural habitat within the range of the San Joaquin kit fox. Thus, there is relatively little spatial overlap between kit foxes and red foxes; however, at least three kit fox deaths have been attributed to red foxes (Ralls and White 1995; Clark et al. 2005). As with coyotes, this mortality appears to be a form of interference competition and not predation. Free-ranging dogs can also be a source of mortality for kit foxes in some areas. These dogs could be free-ranging pets, livestock guarding dogs, or feral dogs. Spiegel (1996) documented a number of kit fox mortalities from dogs in the Lokern area in western Kern County, and a kit fox mortality on the Carrizo Plain was attributed to a domestic dog (Ralls and White 1995).

Deaths of kit foxes by badgers has been assumed, but there is little direct evidence of such predation. The potential seems high because, unlike coyotes and bobcats, badgers are able to easily enter kit fox dens. A badger is suspected of entering a kit fox den and killing a radio-collared kit fox at Camp Roberts (Standley et al. 1992). Great-horned owls also occur throughout much of kit fox range and certainly are sufficiently large and powerful to attack and kill a kit fox. Only one mortality has ever been putatively attributed to a great-horned owl and that was in an urban setting. However, as with golden eagles, this may be a difficult source to detect, because pups may be the primary targets and radio collars have rarely been placed on pups. In addition, the entire carcass may be carried off to a distant location (e.g., eagle or owl nest) and consumed.

Vehicles are usually a secondary source of mortality for kit foxes in natural habitats (Figure 2.10). Generally, this source accounts

FIGURE 2.10. San Joaquin kit fox killed by a vehicle. Photo by Brian Cypher.

for less than 20% of mortalities. Not surprisingly, the bulk of this mortality occurs along paved roads that traverse kit fox population areas. A factor that may help reduce vehicle strikes is the fact that traffic tends to be heaviest during the day when kit foxes are usually resting in their dens. Most road crossings by foxes occur at night when the traffic volumes are lower. In a road effect study conducted in the Lokern Natural Area within the western Kern County core area, 60 kit foxes were radio-collared during 2001–2004. Only one fox was struck and killed by a vehicle, while 21 were killed by larger predators (Cypher et al. 2009). Vehicle mortality rates from other studies were similar and include 2 of 24 collared foxes in western Merced County (Briden et al. 1992), 1 of 41 collared foxes on the Carrizo Plain (Ralls and White 1995), 1 of 103 collared foxes in Lokern (Spiegel 1996), and 20 of 341 (5.8%) collared foxes at Elk Hills (Cypher et al. 2000).

San Joaquin kit foxes occasionally die from exposure to toxins. In particular, rodenticides have been responsible for a number of kit fox deaths. Rodenticides are a particularly significant issue in urban habitats, where the majority of kit foxes have tested positive for exposure to anticoagulant rodenticides (McMillin et al. 2008; Cypher, McMillin, et al. 2014), particularly more toxic second-generation anticoagulant rodenticides such as brodifacoum and bromadiolone, commonly found in products used to control commensal rodents such as Norway and black rats and house mice. At least two kit foxes were documented to have died of rodenticide poisoning in Bakersfield (Cypher, McMillin, et al. 2014).

In natural lands, exposure rates are much lower. For example, no rodenticides were detected in samples from 12 foxes from the Lokern area (McMillin et al. 2008). Outside of urban areas, first-generation anticoagulant rodenticides (e.g., chlorophacinone, diphacinone) are more likely to be an issue, as these are the ones commonly used, legally and illegally, on grazing and agricultural lands to control rodents, particularly California ground squirrels. At least two San Joaquin kit foxes died when rodenticide bait was distributed on private lands adjacent to the Camp Roberts military installation (Standley et al. 1992). Currently, rodenticide-laced baits are still being distributed on many private rangelands to reduce ground squirrel and other rodent populations. It is unclear whether most kit fox poisonings are primary (i.e., direct ingestion of baits) or secondary (i.e., ingestion of rodents and insects killed by rodenticides). Highly toxic second-generation anticoagulant rodenticides are commonly distributed in containers that are difficult for kit foxes to access, and so exposure is likely secondary. First-generation anticoagulant rodenticides are not uncommonly distributed in containers accessible to kit foxes or are simply broadcast on rangelands, and therefore exposure could be either primary or secondary. In general, rodenticides do not appear to be a common or significant source of mortality; however, their effects may also be underestimated. An unknown number of foxes are likely caught and killed by predators or struck by vehicles because they were weakened or incapacitated by rodenticide exposure.

Oil and gas production is extensive in large portions of the range of the San Joaquin kit fox (see Kit Foxes and Oil Fields). Kit foxes readily inhabit oilfields as considerable habitat, including prey, is usually still present (e.g., Spiegel 1996; Cypher et al. 2000). Hazards unique to oilfields are also present, including heavy equipment, toxic substances, open pipes, and open well cellars. Spiegel (1996) reported that a radio-collared fox covered in oil was found alive, but soon died despite efforts to clean and rehabilitate it. Despite these potential hazards, kit fox mortalities due to oilfield activities appear to be rare, particularly with the implementation of required mitigation measures (e.g., capping pipes, rapid cleanup of spills, exclusion fences around hazardous sites) designed to prevent such mortalities.

Other causes of mortality tend to be rather rare, at least in non-urban environments. Despite formal federal and state protections, San Joaquin kit foxes are occasionally shot illegally (Laughrin 1970; Morrell 1972; Briden et al. 1992; Standley et al. 1992; Cypher et al. 2000). Whether people are specifically shooting kit foxes or whether the foxes mistaken for young coyotes is unknown. Foxes can also be entombed when natural lands are converted to other uses. Knapp (1978) documented the entombment and deaths of three radio-collared kit foxes in their dens as the lands on which the dens were located were plowed to plant crops.

Parasites and Diseases

San Joaquin kit foxes, and kit foxes in general, host a number of external and internal parasites, most of which are found on other canids as well and do not appear to cause serious health issues for the foxes. Fleas are the most common ectoparasite found on San Joaquin kit foxes (Figure 2.11). The most frequently found species, *Echidnophaga gallinacea*, also known as the "chicken flea" or stick-tight flea, is cosmopolitan in distribution and commonly found on carnivores (Riner et al. 2018). *Pulex* species and *Ctenocephalides felis* (aka the "cat flea") are typically associated with humans and particularly the dogs and cats they keep as companion animals. These fleas were found on foxes in urban environments and those using solar farms and a military installation (Spencer and Egoscue 1992; Riner et al. 2018). Other flea species found on foxes in non-urban environments have included *Cediopsylla inaequalis*, *Thrassis augustsoni*, *Odontopsyllas dentatus*, *Meringis californicus*, and *Hoplopsyllas anomalus*. These fleas are typically associated with the prey consumed by kit foxes,

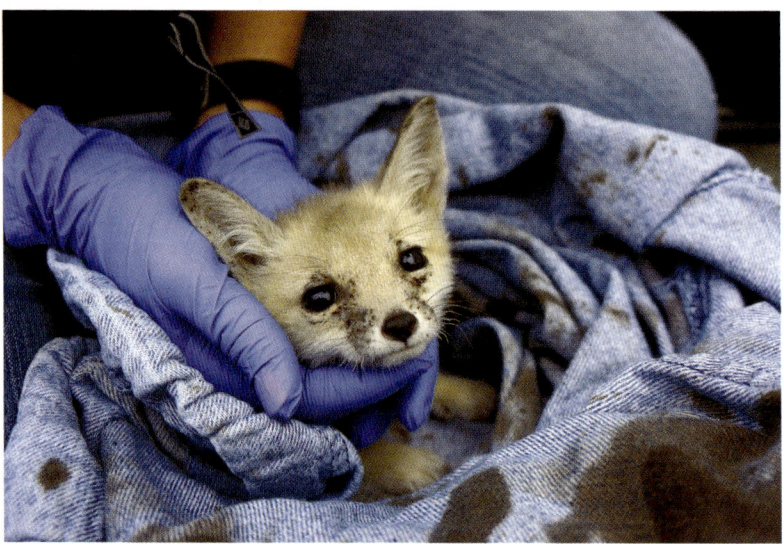

FIGURE 2.11. San Joaquin kit fox with a heavy infestation of fleas. Photo by Erica Kelly.

such as rabbits, squirrels, and kangaroo rats (Egoscue 1985; Spencer and Egoscue 1992; Riner et al. 2018). Ticks have rarely been found on San Joaquin kit foxes, perhaps because the foxes live in an extremely arid environment where it is difficult for ticks to complete their life cycle.

No studies have been conducted to determine whether parasitism by fleas has any measurable effect on kit fox health, either directly through blood consumption or as a vector for pathogens. Perhaps the most noticeable effect may be den switching by foxes (Egoscue 1962; Koopman et al. 1998). Particularly in the early spring when the ground may still be quite damp from winter rains, temperatures are warming, and litters of young pups are in dens, flea populations appear to build to substantial levels (personal observation). Fleas can be seen in abundance around den entrances and, indeed, can be felt bouncing off a hand stuck into the den entrance. Kit fox family groups switch dens relatively frequently at this time. Interestingly, they may move to a den only a short distance away (e.g., <100 m or 325 ft), so clearly, this den switching is not related to predator evasion, as any predator such as a coyote could easily follow the scent trail of the family to the new den. However, a leading hypothesis is that it may result from flea populations becoming intolerable, and so kit foxes move to a new den until flea populations at that den also become intolerable. Good evidence for this theory was provided by Kluever et al. (2019), who found that desert kit foxes treated for fleas used fewer dens overall and used dens longer before switching to a new one.

A number of endoparasites have been found in San Joaquin kit foxes. Similar to the situation with ectoparasites, there is no quantitative information on the effects of these parasites on kit fox health. The most commonly observed endoparasites, probably because they are so conspicuous, are nematodes (roundworms) in the stomach and intestines and cestodes (tapeworms) in the intestines.

Antibodies to a variety of infectious pathogens have been detected in San Joaquin kit foxes. These include brucellosis (*Brucella abortus* and *B. canis*), Cache Valley virus, canine distemper, canine hepatitis virus, canine parvovirus, coccidioidomycosis (*Coccidioides immitis*), Colorado tick fever, leptospirosis, rabies, toxoplasmosis (*Toxoplasma gondii*), tularemia (*Francisella tularensis*), and vesicular stomatitis (McCue and O'Farrell 1988; Standley and McCue 1997; Miller et al. 2000). The detection of antibodies simply means that a fox has been exposed to a disease pathogen but does not indicate active disease. Clearly, kit foxes are being exposed to a number of pathogens, but whether active disease develops from these exposures and the potential individual and population effects from the pathogens are unknown.

Similar to parasites, diseases have generally not been a significant issue for kit foxes. Indeed, disease in kit foxes hardly warranted mention before about 2000. Prior to this, no occurrences of population effects due to disease had been reported for kit foxes; however, San Joaquin kit fox numbers in the Salinas Valley, particularly near Camp Roberts in San Luis Obispo/Monterey Counties, declined significantly in the 1990s (White et al. 2000). Several factors, or potentially a combination of factors, may have been responsible, but rabies was hypothesized as one potential cause of this decline. A rabies epidemic occurred regionally in the striped skunk population during this time, and this may have spilled over into kit foxes. Two kit foxes with rabies were documented at Camp Roberts during this time (Standley et al. 1992), as was a rabid skunk (White et al. 2000). Also, kit foxes and skunks are known to use the same dens, including simultaneously on occasion, and they frequently use common food sources in urban environments (Harrison et al. 2011). Thus, the potential for interspecific transmission of disease between these two species is high. Rabies is always fatal to infected foxes, and this disease has caused significant population declines in a variety of canid species worldwide (Laurenson et al. 2004). It is also a zoonotic disease that can potentially be transmitted to humans, which enhances its importance.

Canine distemper virus is another viral disease of concern. Distemper has caused significant mortality in gray foxes, island foxes, and red foxes in North America (Cypher 2003). Distemper almost caused the extirpation of the island fox subspecies inhabiting Catalina Island (Coonan et al. 2010). In 2011, dead and sick desert kit foxes were reported near a solar farm under construction near Blythe in the Mojave Desert in California. The cause was determined to be distemper. At least 10 foxes were documented to have succumbed to the disease (Clifford and Rudd, CDFW, unpublished data). This appeared to be a relatively localized event, as kit foxes being monitored at nearby sites did not appear affected. In 2019, three kit foxes found dead near a solar site in the Panoche Valley tested positive for distemper, although it was unclear whether distemper caused the deaths (ESRP, unpublished data); however, vigilance is warranted, given the potential for distemper to adversely affect canid populations (Laurenson et al. 2004), and also given the potential for transmission from domestic dogs. Distemper antibodies are consistently detected in San Joaquin kit fox populations, and McCue and O'Farrell (1988) suggested that distemper may be enzootic in the populations.

More recently, significant impacts to San Joaquin kit fox populations from disease were documented. In spring 2013, sarcoptic mange was detected in San Joaquin kit foxes in Bakersfield (Cypher et al. 2017). Prior to this, sarcoptic mange had never been documented in kit foxes. Sarcoptic mange is caused by a mite, *Sarcoptes scabiei*, that burrows into the skin of animals. The animals react to this burrowing and to the saliva and excrement of the mites, and characteristic symptoms include intense pruritus (itching) and dermatitis, alopecia (hair loss), hyperkeratosis (thickening of the skin), and encrustations (crusty dried exudates), secondary bacterial infections, and finally extreme morbidity and death (Figure 2.12). In essence, mange is a highly transmissible ectoparasitic infection with

FIGURE 2.12. San Joaquin kit foxes with sarcoptic mange. Photos by Tory Westall.

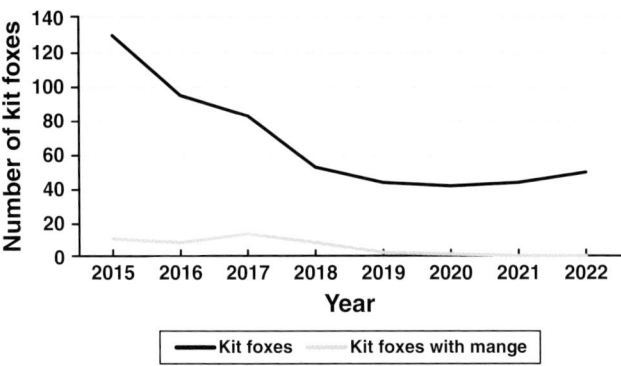

FIGURE 2.13. Number of individual San Joaquin kit foxes observed during systematic city-wide camera station surveys in Bakersfield, CA from 2015 to 2022.

serious health consequences. Mange affects a large number of species (Pence and Ueckermann 2002), and in some species (e.g., coyotes, red foxes) some individuals apparently survive. However, in kit foxes, all cases apparently result in death unless the individual is treated (Cypher et al. 2017).

Mange rapidly spread throughout the Bakersfield kit fox population. Beginning with the first detection in 2013, efforts were initiated to respond to reports of sick foxes and attempt to capture them to provide medical care; however, the number of foxes affected rapidly became too great, and the area over which cases were occurring became too large. Annual monitoring of the Bakersfield kit fox population using systematic camera station surveys was initiated in 2015 and revealed a substantial population decline (Figure 2.13) that probably began in 2013 and continued until at least 2019.

Significant effort has been expended by staff of the Endangered Species Recovery Program, California Department of Fish and Wildlife, and various volunteers to capture and treat kit foxes with mange. As of early 2023, 141 foxes with mild cases of mange have been treated in the field, and 155 foxes with more severe cases have been treated at the wildlife rehabilitation clinic at the California Living Museum (CALM) in Bakersfield (see Research and Conservation Needs). Field treatment consists of applying a topical dose of selamectin (aka Revolution, the same product used to protect domestic cats and dogs from fleas and ticks). Treatment at CALM consists of three applica-

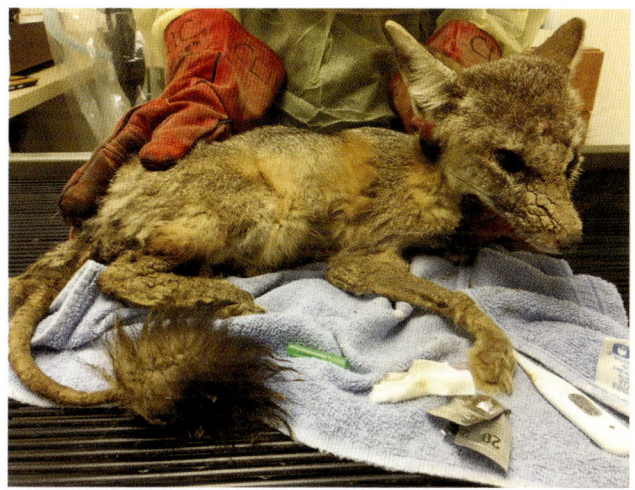

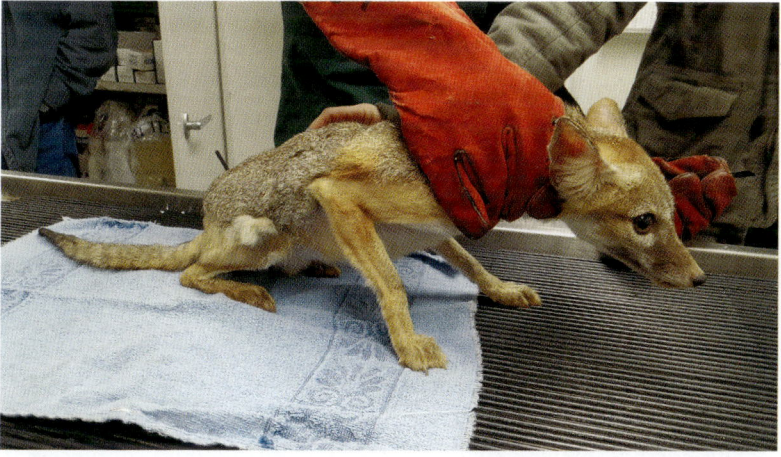

FIGURE 2.14. A San Joaquin kit fox before and after it was treated for sarcoptic mange at the California Living Museum. Photos by Erica Kelly.

tions of selamectin, with the first dose applied when a fox is admitted to the clinic and then subsequent doses applied two and four weeks later (Figure 2.14). Other supportive care includes fluids, antibiotics to treat secondary infections, removal of dead skin and loose hair, and heating pads and lamps if hair loss is severe. Despite these efforts, foxes with advanced cases of mange sometimes do not survive. Once a fox is considered recovered, it is released back at its original capture site.

The number of cases of mange among Bakersfield kit foxes has declined dramatically, and no cases were detected during the summer 2022 survey. Time will tell whether the epidemic has subsided and ended, or whether mange in the Bakersfield kit fox population might become endemic, as has been observed in some red fox, coyote, and gray wolf populations (Niedringhaus et al. 2019). However, there is cautious optimism that even if mange does become endemic in this population, it will not drive the population to extinction.

It is unclear how mange entered the Bakersfield kit fox population. As stated previously, mange had not been documented in kit foxes prior to its appearance in the Bakersfield population. Mange was never detected among the hundreds of kit foxes that have been captured for various studies or detected on automated camera stations or collected as roadkills. Thus, it seems that the origin was not other kit fox populations; however, coyotes with mange had been observed for about a decade prior all around the San Joaquin Valley, and indeed in many other parts of California. Mange had also been observed in red foxes in various locations around the state. Red foxes do occur in Bakersfield, but none with mange had been observed prior to its detection in kit foxes. Mange can also occur in other wild mammals that occur in Bakersfield such as striped skunks, raccoons, opossums, and feral cats; however, mange had not been detected in any other species in Bakersfield prior to the kit fox cases. Finally, it is possible that a domestic dog with mange had been brought to Bakersfield and that mange somehow was transferred to a kit fox.

The most plausible explanation for the introduction of mange is that it likely spilled over from one of the many coyotes with mange observed around Bakersfield. However, there is a conundrum here in that when a coyote comes close enough to a kit fox to transfer mites, the coyote usually kills the fox. It is possible that a kit fox rested in a location where a coyote rested, but this would have been a considerable coincidence. Images captured on an automated camera station at a site just east of Bakersfield suggested another mode of transmission. In these images, a coyote with mange was seen extending its head and shoulders well into the entrance of a kit fox den. The coyote was making contact with the sides of the den entrance, and mites could have dropped off the coyote onto the soil. Mites can live away from their hosts for a period of time. Additional camera station images showed kit foxes sitting at the den entrance just about an hour after the coyote visit, and the foxes could have easily picked up mites from the contaminated soil. Coyotes commonly explore kit fox dens, particularly when young pups are present. Thus, this is a very plausible mode of mite transmission from coyotes to kit foxes.

The mode of transmission among foxes is also not clear-cut. Mange mites are typically transferred between individuals by direct contact. Kit foxes do have frequent contact with other members of their social group. In particular, they share dens and likely lie in close proximity within the dens. However, based on numerous observations, kit foxes appear to rarely have physical contact with individuals outside their social group. This is where urban living and dens may play a significant role. Kit fox density is considerably higher in urban environments compared with natural habitats (Cypher, Deatherage, et al. 2023; see Kit Foxes and Urban Areas). Consequently, spatial overlap among individuals and among social groups is high. Because of this greater spatial overlap, overlapping den use is also higher. Foxes from different social groups do not typically occupy dens simultaneously; however, Loredo et al. (2020) estimated that climatic conditions in kit fox dens were sufficient to allow mites to survive off-host potentially for up to 7 days. Thus, off-host survival of mites would provide considerable opportunity for a fox with mange to use a den and contaminate it with mites that would then infest another fox that subsequently used the same den.

Unfortunately, in winter 2019, kit foxes with mange began appearing in the small town of Taft, located about 35 km (22 mi) west of Bakersfield. Similar to Bakersfield, the density of kit foxes in Taft is considerably higher than that in non-urban populations. The route by which mange entered the Taft population is unknown. A fox or foxes potentially could have traveled from Bakersfield to Taft; however, that would constitute a considerable distance to travel even for a strong, healthy fox without mange. Also, no foxes with mange have been detected in the area between Bakersfield and Taft. The introduction of mange in Taft could have also resulted from another spillover event, as coyotes are abundant in the lands around Taft, and individuals with mange have been observed. Similar to Bakersfield, efforts have been conducted to capture and treat foxes in Taft.

In a camera station survey conducted in fall 2022, no kit foxes with mange were detected.

Encouragingly, mange has not been detected in non-urban kit foxes. Monitoring with camera stations was initiated in areas just outside of Bakersfield and Taft in 2017. As of the 2022 surveys, no kit foxes with mange had been detected in these areas. Also, trapping efforts for various research projects and camera station surveys have been conducted at numerous locations throughout the range, and many kit fox carcasses (primarily roadkills) have been examined. Higher fox population density and greater spatial overlap resulting in higher rates of use of common dens appear to be critical factors in facilitating the transmission of mange mites throughout a population. Fox densities and spatial overlap are lower in non-urban kit fox populations, and consequently mange has not been detected in these populations to date despite extensive surveys and monitoring.

REPRODUCTION

Reproductive success is the other main demographic parameter that, in addition to survival, is a primary factor driving population dynamics. Other than the annual timing of different reproductive events, reproduction by San Joaquin kit foxes is quite similar to that of other foxes. San Joaquin kit foxes that are not already paired with a mate begin to do so in about November. Exactly how the foxes find potential mates and select one is unknown. At this time of year, testes on the males have begun to enlarge, and the vulvas on females begin to swell as they enter proestrus. In early to mid-December, the females become fully estrus and are receptive to attempts by males to mate. At this point, the testes on the males are fully descended and swollen, and the scrotum may exhibit patches of thin hair or alopecia, which helps to cool the testes and enhance sperm production. Mating is typical of that of canids with the male mounting the female from behind, and then after ejaculating, the two become locked in a copulatory tie for some minutes. After a gestation of approximately 51–54 days (Zoellick et al. 1987), litters are born, usually around mid-February; however, litters can be born as early as mid to late January or as late as mid-March.

Reproduction by most mammals is energetically expensive, par-

FIGURE 2.15. Male San Joaquin kit fox bringing four kangaroo rats to a natal den. Photo by Tory Westall.

ticularly late gestation, lactation, and provisioning of young until they are able to forage independently. Therefore, the timing of these events is evolutionarily selected to coincide with periods of resource abundance (Ewer 1973). This is true for kit foxes as well. Abundance of their primary prey, kangaroo rats, is highest during late winter-early spring. This was demonstrated well by Morrell (1971), who indexed kangaroo rats one or more times each month for most months of the year and found that kangaroo rat abundance was highest during March–May. This is precisely when San Joaquin kit foxes are weaning their pups and the pups are growing rapidly and require large quantities of food (Figure 2.15).

Typical of canids, the neonates are blind and naked. Around the time of parturition, the female may remain in the den for several days without emerging. Her mate will bring food to the den for her during this time. When the pups are approximately 3–4 weeks of age, their eyes are open, and they begin to appear at the entrance of the den. By about mid-March, most litters have emerged from their natal dens (Figure 2.16). As they age, the pups spend more time outside the

FIGURE 2.16. San Joaquin kit fox pups 4–5 weeks old, soon after emerging aboveground. Photo by Tory Westall.

den and slowly expand their area of activity. As activity increases, the areas around natal dens exhibit lots of fresh digging, and the vegetation around natal dens tends to become trampled to the extent that natal dens are even identifiable from the air. The weaning process begins at around 5–6 weeks of age, and the pups are usually fully weaned by 8 weeks. At this point, prey remains become more common around natal dens. The pups become more and more active and also exhibit an increasing amount of diurnal activity in their zest to play and explore. The pups also begin to hunt and capture insects around the den.

Beginning about mid to late April, when the pups are 8–10 weeks old (Figure 2.17), the parents may actually spend more time in other dens than the one the pups are in (but nearby), probably both to get some rest and also because it likely becomes increasingly crowded in the natal den as the pups rapidly increase in size. During the pup-rearing period, the adults may also move the pups to a new den. Usually the new den is relatively close to the previous den (e.g., <100 m), and these moves may be related to flea infestations reaching intolerable levels in the old den (see Survival and Mortality Factors). Usually by June, the pups are becoming quite independent and may

FIGURE 2.17. San Joaquin kit fox pups about 10–12 weeks old. Photo by Christine Van Horn Job.

follow the parents on foraging trips. Not uncommonly by this time, members of a litter may be split between two or more dens. In June, some pups even begin to disperse from their natal ranges (see Space Use, Movements, Dispersal, and Activity).

Reproductive success in kit foxes is commonly based on the presence of pups at natal dens. Success rates reported from various studies range from 0% to 100% (Table 3). As with many species, reproductive success is strongly influenced by food availability. Thus, in years of low prey availability, success is usually lower. In really poor years, reproduction can fail completely, with few or no foxes successfully producing litters. Reproductive failure in years of low food availability was documented on two study sites on the Carrizo Plain (White and Ralls 1993; Cypher, Fiehler, et al. 2014) and at the Naval Petroleum Reserves (Cyphcr et al. 2000).

It is not clear whether reproductive failure is more common at a particular stage in the reproductive process. Energetically, ovulation and mating are not terribly costly (Oftedal and Gittleman 1989). Gestation is energetically more expensive as resources are needed at an increasing rate by the growing fetuses. So, this is one potential point of failure if sufficient resources are not available. Zoellick et al. (1987)

estimated that productivity loss among adult female foxes from ovulation to 10 weeks post-parturition was 19%. Among mammals, lactation is considered the most energetically taxing period in the reproductive process (e.g., Gittleman and Thompson 1988; Oftedal and Gittleman 1989). The female must consume enough food to produce the milk required by the growing pups. This may be the most common point of reproductive failure during years of low food availability. Finally, sufficient prey obviously needs to be available to nourish the growing pups once they are weaned.

Another obvious cause of reproductive failure is the death of the female while the pups are still nursing. Once the pups have begun the weaning process, either of the adults can continue to raise the litter in the absence of the other adult. Sometimes, one or more "helpers" may be present (see Social Ecology), and this greatly facilitates the rearing of pups by one parent.

It is unclear whether factors other than nutrition (e.g., female age) influence litter size in kit foxes. Mean litter size for San Joaquin kit foxes is approximately four, although litters as large as nine pups have been documented (Table 3). Sex ratios within the litters tend to be about even (e.g., Cypher et al. 2000). However, male-biased sex ratios among pups have been reported from areas where food availability was low or declining and kit fox abundance was declining as well (Egoscue 1975; Zoellick et al. 1987). Male-biased ratios among young are not uncommon in populations of mammals that are nutritionally stressed (McGinley 1984). Males are more likely to disperse and therefore find better conditions elsewhere, and by departing they are reducing competition for resources in the natal area. As with dogs, observations of "runts" in litters have been documented. Such runts seem to fare fine as long as food is plentiful.

POPULATION DYNAMICS

San Joaquin kit fox populations can fluctuate markedly among years (e.g., Cypher et al. 2000). These fluctuations are attributable to variation in food availability resulting from the variable conditions in the arid environments in which kit foxes live. Environmental conditions in arid environments are inextricably linked to precipitation patterns, and conditions can change significantly with even small differences in the amount of precipitation received each year (Frank and Inouye 1994; Previtali et al. 2009). Even just differences in the timing and pattern of precipitation can seriously alter conditions between years, even if the precipitation totals are similar. In short, the amount and pattern of precipitation falling each year influences primary productivity, which in turn affects the availability of prey, particularly small rodents but also invertebrates, birds, lizards, rabbits, and other items (Ostfeld and Keesing 2000; Germano et al. 2012; Prugh et al. 2018). Annual prey availability affects kit fox survival and reproduction, and these processes determine whether kit fox abundance increases, decreases, or remains the same between years.

The relationship between precipitation, prey abundance, and fox abundance is complex, and thus, predicting kit fox population trends is difficult. Responses by prey and fox populations can exhibit both cumulative and lag or delayed effects. For example, many prey populations, particularly kangaroo rats, which are the main prey of kit foxes, can usually weather one year of below average precipitation relatively well without a substantial decline in abundance (Germano and Saslaw 2017). However, two or more consecutive years of low rainfall usually cause significant declines in numbers of kangaroo rats, rabbits, and other species. Conversely, particularly following such periods of low rainfall, a year with above average or even average precipitation will result in an increase in the numbers of prey. Two or more consecutive years of good precipitation, even if the amounts are the same each year, can result in prey numbers continuing to increase until other factors (e.g., social interactions, disease, etc.) inhibit or prevent further increases (Greenville et al. 2012; Germano and Saslaw 2017).

Also, because of differences in life history strategies and competitive interactions, different species may exhibit asynchronous trends. For example, because giant kangaroo rats store large quantities of food in underground larders (Shaw 1934), they may not decline as quickly during droughts compared with smaller kangaroo rats that store only small quantities of food in surface caches (Best 1991; Reichman and Price 1993). Also, larger kangaroo rats can competitively exclude smaller kangaroo rats (Figure 2.18), and both can in turn exclude smaller species such as pocket mice (Brown and Harney 1993; Tennant and Germano 2013). Short-nosed kangaroo rats

TABLE 3. *Reproductive success and litter size for San Joaquin kit foxes from locations throughout the range*

Location	Year(s)	No. females	Mean % success	No. litters	Mean litter size	Range	Source	Notes
Western Merced Co.	1985–87	—	—	7	2.4	1–3	Briden et al. 1992	Satellite population area; natural lands
Northern Carrizo Plain, San Luis Obispo Co.	2013	3	0.0[a]	—	—	—	Cypher, Fiehler, et al. 2014	Core population area; natural lands
Carrizo Plain, eastern San Luis Obispo Co.	1989–91	19	21.1[b]	4	2.0	1–3	White and Ralls 1993	Core population area; natural lands; low prey availability due to drought; 0% success in 89 and 90
Camp Roberts, northern San Luis Obispo Co.	1989–91	38	31.6[b]	21	3.0	1–6	Spencer et al. 1992	Satellite population area; natural lands with occasional military training activities
Carrizo Plain, San Luis Obispo Co.	2016	9	44.4[a]	4	4.0	3–5	Cypher, Westall, et al. 2022	Core population area; natural lands; low prey availability
Lokern Natural Area, western Kern Co.	1990–93	12	50.0	6	4.2	3–5	Spiegel 1996	Core population area; oil and gas production
Elk Hills, western Kern Co.	1980–86	37	51.0	25	4.1	—	Cypher et al. 2000	Core population area; natural lands
Lokern Natural Area, western Kern Co.	2002–04	24	58.3	23	3.8	2–9	Cypher et al. 2009	Core population area; natural lands
Lokern Natural Area, western Kern Co.	1990–93	25	64.0	14	3.7	2–5	Spiegel 1996	Core population area; natural lands
Elk Hills, western Kern Co.	1980–86	32	69.0	27	4.4	—	Cypher et al. 2000	Core population area; oil and gas production
Carrizo Plain, San Luis Obispo Co.	2018	11	72.7[a]	8	4.4	2–7	Cypher, Westall, et al. 2022	Core population area; natural lands; high prey availability

Location	Year(s)	No. females	Mean % success	No. litters	Mean litter size	Range	Source	Notes
Panoche Valley, eastern San Benito Co.	2020–22	16	81.3	23	4.0	1–7	Cypher, Westall, et al. 2023	Core population area; natural lands; reference site for Panoche Valley Solar Farm study Year 1; high prey availability
Bakersfield, Kern Co.	1997–2004	78	82.4	91	3.8	1–9	Cypher 2010; unpubl. data	Satellite population area; urban
California Valley Solar Ranch, San Luis Obispo Co.	2015–17	15	86.7	7	4.5	2–7	HTH 2019	Core population area; natural lands; reference site for California Valley Solar Ranch study
California Valley Solar Ranch, San Luis Obispo Co.	2015–17	15	86.7	11	3.2	1–5	HTH 2019	Core population area; solar site
Topaz Solar Farms, San Luis Obispo Co.	2015–17	9	88.9	10	3.9	1–7	Cypher, Westall, et al. 2019	Core population area; natural lands; reference site for Topaz Solar Farm study
Panoche Valley, eastern San Benito Co.	2020–22	14	92.9	12	3.4	1–5	Cypher, Westall, et al. 2023	Core population area; Solar site Year 1; high prey availability
Topaz Solar Farms, San Luis Obispo Co.	2015–17	10	100.0	12	4.3	2–8	Cypher, Westall, et al. 2019	Core population area; solar site
Panoche Valley, eastern San Benito Co.	2015–16	2	100.0	2	4.0	3–5	Cypher et al. unpubl. data	Core population area; natural lands; pre-solar farm construction; low prey availability

[a] One-year study.
[b] Percent success is for all years combined.

Notes: The "% success" is the proportion of females for which pups were verified surviving to emergence from natal dens. Data are ordered from lowest to highest Mean % success.

FIGURE 2.18. Three kangaroo rat species commonly consumed by San Joaquin kit foxes: giant kangaroo rat (left), Heermann's kangaroo rat (middle), and short-nosed kangaroo rat (right). Photo by Brian Cypher.

increased on the Carrizo Plain as larger giant kangaroo rats declined (Prugh et al. 2018), and San Joaquin pocket mice increased at Elk Hills when kangaroo rat abundance declined (Cypher 2001).

Finally, to further complicate matters, it is actually possible for conditions in the San Joaquin Desert to be "too good." In particular, precipitation exceeding the annual average results in high plant productivity and very dense ground cover, especially of non-native grasses. Kit foxes and their prey, and indeed many co-occurring animals and plants, are adapted to arid conditions, including sparse ground cover, usually with abundant bare ground. Thus, the prey species in particular have a difficult time dealing with dense cover. Their mobility is reduced and predation risk increases; consequently, their numbers may actually decline under these circumstances. A prime example occurred in 1997–1998 when record high precipitation was recorded in the San Joaquin Valley following several years of above average precipitation. Vegetation became extremely dense. Furthermore, frequent fog and cooler temperatures kept this vegetation wet for extended periods. Kangaroo rats captured during this time exhibited signs of illness, including coughing and respiratory distress. Kangaroo rat numbers crashed to low levels by that spring (Single et al. 1996), and the lack of food resulted in a decrease in kit fox numbers as well in the Carrizo Plain (CDFW unpublished data) and western Kern County core areas (Cypher et al. 2000). Interestingly, a critical factor in this scenario was the abundance of non-native plants, particularly Mediterranean grasses like red brome, that exacerbate the situation (Single et al. 1996; Germano et al. 2001).

As a consequence of the complex ecological processes described above, there are typically not strong direct correlations between annual precipitation, most prey populations, and kit fox population trends. However, in general when annual precipitation is higher, overall prey availability increases, kit foxes exhibit higher survival and reproductive rates, and their numbers increase. In drier years, foods become less available, which results in low or no reproduction among kit foxes, and their numbers decline as well (White and Garrott 1999; Cypher et al. 2000). As a result of the seasonal timing of the ecological events involved in kit fox population dynamics, typically there is a lag effect. For example, when adequate winter-spring precipitation finally returns after multiple years of drought conditions, by the time plants grow and prey species begin to increase in abundance, kit fox reproduction has usually already failed for that year. Thus, a significant increase in pup production and kit fox numbers may not be observed until the following reproductive year, assuming the improved conditions continue (White and Garrott 1997; Cypher et al. 2000). Similarly, as mentioned above, kit fox prey species, particularly kangaroo rats which store food, can typically weather one year of poor conditions before their numbers begin to decline. Thus, it may take two or more consecutive years of poor conditions before prey numbers become sufficiently low that kit foxes cannot successfully reproduce. Kit foxes are also sufficiently adaptable that they can switch to using alternate foods if their preferred prey, usually kangaroo rats (see Foraging Ecology), become scarce. This was observed at Elk Hills, where jackrabbits constituted the primary prey for kit foxes in the early 1980s when kangaroo rat abundance was low (Cypher et al. 2000). This prey switching can help to delay and moderate kit fox declines.

Precipitation is the best predictor of kit fox population trends, simply because it generally correlates with overall prey availability,

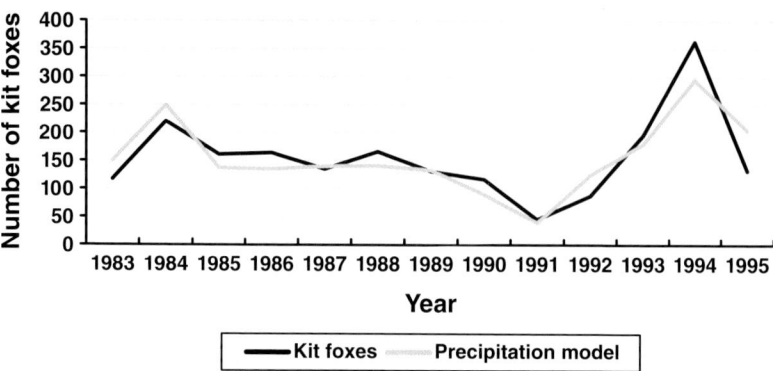

FIGURE 2.19. Number of San Joaquin kit foxes by year relative to the number predicted based on a precipitation model (see Cypher et al. 2000) for the Elk Hills in Kern County, California.

even if it is not highly correlated with the annual availability of any given prey species. At both Elk Hills and the Elkhorn Plain area of the Carrizo core area, kit fox population trends mirrored annual precipitation reasonably well with obvious lag effects of 1–2 years. At Elk Hills, total rodent abundance and a composite prey index for rodents and rabbits combined were both significantly correlated with annual precipitation, although because of the lag effects discussed previously, the correlation was with the precipitation total from the previous year (Cypher et al. 2000). Kit fox density was also related to the annual composite prey index. Interestingly, modeling revealed a very strong correlation between kit fox abundance and a combination of annual precipitation from the three previous years (Figure 2.19; Cypher et al. 2000). This nicely exemplifies the lag and cumulative aspects of the relationship between kit fox abundance, prey availability, and precipitation.

Variability in the annual availability of potential kit fox prey can be quite dramatic. On a standard small mammal monitoring grid (144 traps) in the Lokern area, the number of individual giant kangaroo rats captured annually ranged from 0 to 172 over a 13-year period (Germano and Saslaw 2017). Giant kangaroo rat numbers on monitoring plots on the Carrizo Plain declined from more than 2100 in 2011 to fewer than 200 in 2014 in response to drought conditions (Prugh et al. 2018), and then their numbers increased to 3009 by 2017 as conditions improved again (Bean et al. 2020). Similar dramatic fluctuations are observed among other potential prey species such as rabbits, squirrels, and invertebrates (Cypher et al. 2000; Germano et al. 2012; Prugh et al. 2018).

This variation in food availability can produce fluctuations in kit fox abundance that not only are marked but can also occur relatively rapidly. For example, during a 15-year period (1981–1995) in which kit fox abundance was monitored at the Naval Petroleum Reserves in Kern County, the lowest (1991, 46 foxes) and the highest (1994, 363 foxes) population estimates were recorded just 3 years apart (Cypher et al. 2000). The lowest estimate followed a multi-year drought, and the highest estimate followed three consecutive years of average or above average precipitation. Marked fluctuations were also evident in the 40-year data set collected by the California Department of Fish and Wildlife on the Elkhorn Plain.

Several important points related to these observed population dynamics among kit foxes stand out. The first is that such dynamics are typical of organisms in arid or desert environments. Their reproductive potential is such that they are able to rapidly take advantage of "boom" years when resources are abundant (e.g., Ostfeld and Keesing 2000; Greenville et al. 2012). In the case of kit foxes, more females successfully reproduce, litter sizes are larger, and juvenile survival is higher. Thus, populations increase. The second point is that these dynamics also have potentially dangerous consequences for San Joaquin kit foxes as well as other rare species whose habitat has been reduced and fragmented. When the populations decline to low levels during periods of low resource availability, the risk of extinction of small populations in habitat fragments increases (Frankham et al. 2017). These extirpations can result from demographic stochasticity or catastrophic events. For example, a fluctuating population of wild dogs in the Serengeti went extinct from chance factors when the markedly fluctuating population was at low numbers (Ginsberg et al. 1995).

A third point is that the population dynamics of San Joaquin kit foxes are driven by prey availability, which in turn is driven by primary productivity, which in turn is a function of annual precipitation patterns. Thus, this constitutes a "bottom up" process. When kit fox abundance increases, the abundance of their competitors likely increases as well, because they also benefit from the increased prey availability. However, these higher levels of competitors are not what limit or drive kit fox populations; instead, it is food availability (White and Garrott 1997, 1999; Cypher et al. 2000). As mentioned, a reduction in coyote abundance at the Naval Petroleum Reserves did not result in an increase in kit fox abundance (Cypher and Scrivner 1992).

Kit fox population dynamics may be more pronounced on the margins of the range. Habitat conditions tend to be less optimal in these marginal areas; in particular, vegetation density tends to be higher. Thus, even years with average precipitation can result in vegetation density that is too high for kit foxes and their prey species. Consequently, in some marginal areas, kit foxes may be at low numbers or even not present in many years. Interestingly, kit foxes tend to occur more frequently and in greater numbers in some of these marginal areas during years of below average precipitation, when the lower vegetation density is more favorable for kit foxes and their prey.

The actual demographic attributes producing changes in kit fox abundance seem to consist of a combination of reproductive and survival processes. As mentioned, in years with good conditions, reproductive output increases, resulting in more pups. White and Garrott (1999) suggested that juvenile survival was the demographic attribute most correlated with changes in population size between years. This assessment was based on survival of pups once they emerged from the den or became independent. However, it is likely more complex than this, as in really poor years, no pups are observed with many females, and the females do not appear to have nursed (i.e., no enlarged, darkened teats). Thus, litters apparently are being lost *in utero*. Interestingly, adult foxes may fare just fine during these lean times. In one study on the northern Carrizo Plain during a period of drought (Cypher, Fiehler, et al. 2014), no reproduction was documented in spring 2013. However, the adults captured on the site trended toward higher weights, and their body condition was excellent. During this time, they were consuming primarily invertebrates. Invertebrates (e.g., insects, scorpions) have high nutritional value, and the adults were able to sustain themselves adequately on this diet; however, this nutrition apparently was not sufficient to complete the energetically costly task of nourishing fetuses or nursing young pups.

Because of the significant annual variation in environmental conditions and the resulting fluctuations in kit fox abundance, identifying a typical density of kit foxes is difficult. As an example of this, White et al. (1996) reported that kit fox density on a Carrizo Plain study area declined from 0.24/km^2 (0.09/mi^2) in 1989 to 0.15/km^2 (0.06/mi^2) in 1991 during drought conditions. At Camp Roberts during the drought, density declined from 1.0/km^2 (0.39/mi^2) in 1988 to 0.5/km^2 (0.19/mi^2) in 1990 (Berry and Standley 1992). After the drought, Spiegel (1996) estimated density at 1.2/km^2 (0.46/mi^2) in the Lokern area, although this estimate is probably high because the study site was small, and the estimate likely included foxes that were using only a portion of the site. In a striking example of the effects of environmental variation, the kit fox population at Elk Hills was monitored from 1981 to 1995. Annual density based on mark-recapture estimates ranged from 0.21 to 1.68/km^2 (0.08–0.65/mi^2), with the 15-year mean being 0.82/km^2 (0.32/mi^2) (Cypher et al. 2000). In addition to this marked variation in density among years, density also varies within each year in association with life history events. In a given year, density is highest in spring with the presence of young of the year, and then density declines during the year because of mortalities until new litters are produced the following spring. Finally, density also varies with habitat quality. In areas with lower habitat quality such as some satellite population areas, kit fox home ranges are typically larger (see Space Use and Movements), meaning that there are fewer foxes per unit area (i.e., lower density).

FORAGING ECOLOGY

Foraging ecology, sometimes simply referred to as "food habits", is a standard attribute of the biology of any animal species. Although commonly viewed as just another detail in the life history of a species, in many respects foraging ecology may be the most important attribute. All life requires a source of energy, without which it cannot exist. For animals, foraging ecology is the study of how they obtain that energy. Furthermore, foraging ecology details the role of a spe-

FIGURE 2.20. Giant kangaroo rat. Photo by Brian Cypher.

cies within a food or energy web and helps define the interactions of the species with the biotic and abiotic constituents of the ecosystem in which the species occurs. Foraging ecology is shaped by a complex interplay between the costs and benefits of feeding on a given item (Stephens et al. 2007). Potential costs include search time, pursuit and capture effort, handling and consumption effort, digestibility, presence of any toxins, potential exposure to interspecific and intraspecific competitors, and other factors. The goal in foraging is to maximize the number of calories obtained while minimizing the costs of obtaining those calories. Hence, this goal gives rise to the term "optimal foraging strategy". In that context, we can consider the foraging ecology of the San Joaquin kit fox.

Most foxes tend to be omnivorous and therefore consume a variety of animal and plant food items. Kit foxes (and indeed, other arid land foxes as well) consume primarily animal foods, largely because plant foods, fleshy fruits in particular, tend to be rare in many arid environments. The main types of foods consumed by kit foxes are rodents, lagomorphs, and invertebrates, supplemented with occasional birds and reptiles. The exact species consumed typically

FIGURE 2.21. Giant kangaroo rat precincts (burrow systems) on the Carrizo Plain National Monument, California. The precincts are the many dark green circular areas in the image. Photo by Brian Cypher.

varies with the local availability of species. Usually, rodents are the primary items consumed by kit foxes.

Although kit foxes will consume a variety of rodents, they are essentially kangaroo rat specialists (Figure 2.20). Kangaroo rats are desert-adapted animals and are typically abundant in arid environments. When conditions are good, numbers can truly be large. Giant kangaroo rats, which are a primary food for kit foxes in all core areas, have been recorded achieving densities of 110 animals per ha (44/ac; USFWS 1998). A stunning visual example of the abundance of giant kangaroo rats can be found in the Carrizo Plain core area, where the precincts (burrow systems) of this species cover the landscape (Figure 2.21).

Kangaroo rats use ricochetal locomotion and can move quickly; however, kit foxes are immensely skilled at catching kangaroo rats. Few observations of kit foxes hunting and capturing kangaroo rats are available; however, kit foxes seem to cruise around searching for food and likely surprise kangaroo rats as they emerge from their burrows to forage. Kangaroo rats do not venture far from their burrows, so the pursuit is usually short. Kit foxes are capable of extremely quick dashes and likely either capture a kangaroo rat in a second or two, or the kangaroo rat makes it back into its burrow and the kit fox likely moves on and continues searching. The efficiency with which kit foxes can prey on kangaroo rats is evident by the images of kit foxes returning to natal dens with three or four kangaroo rats in their mouth within just an hour or so of having begun foraging (see Figure 2.15).

Another reason that kangaroo rats may constitute the primary prey for kit foxes is that in most locations, at least two and sometimes more kangaroo rat species co-occur. This overlap not only increases the amount of food potentially available for kit foxes, but it may also enhance the stability of the amount of preferred prey available. Annual environmental conditions in arid environments fluctuate markedly, and different kangaroo rat species respond to these fluctuations in slightly different ways (see Population Dynamics). Thus, their population trends are not usually synchronous. This benefits kit foxes, because if one kangaroo rat species declines in abundance, one or more other species may still be sufficiently abundant to maintain the fox population in a given area. In most locations, kit foxes likely focus foraging efforts on areas where kangaroo rats are abundant. However, they are undoubtedly opportunistic and consume other food items they encounter while hunting for kangaroo rats. This flexibility in diet would further enhance their foraging efficiency.

Kit foxes also consume rabbits, primarily black-tailed jackrabbits and desert cottontails (Cypher 2003). These are larger meals with a higher energetic reward compared with kangaroo rats; however, there are costs and challenges in hunting rabbits. They occur at lower densities than kangaroo rats and therefore are more difficult to find. Rabbits are fast runners and therefore more challenging to pursue and capture. Jackrabbits are structurally similar in size and weight to a kit fox, and there is some risk to foxes in hunting them. Egoscue (1962) reported observations of kit foxes having difficulty in subduing adult jackrabbits and speculated that some foxes may be seriously injured in attempting to do so. Thus, rabbits may constitute more of a supplemental food item that is opportunistically captured; however, when kangaroo rat availability is low, kit foxes may more actively hunt for rabbits, and their importance in the diet can increase considerably. Rabbits were the primary prey of kit foxes at Elk Hills in the early 1980s when regional kangaroo rat abundance was low (Warrick and Cypher 1999; Cypher et al. 2000).

As with rabbits, squirrels are probably taken more opportunistically. The species in kit fox diets are primarily California ground squirrels and San Joaquin antelope squirrels. These species occur at lower densities than kangaroo rats, and their distributions tend to be patchy. Also, ground squirrels are diurnal, whereas kit foxes are primarily nocturnal. All of that said, California ground squirrels can make up a significant and even primary item in the diet of kit foxes in marginal areas of the range. These areas tend to have denser vegetation (part of what makes them marginal habitat) more favorable to ground squirrels and less favorable to kangaroo rats. California ground squirrels were the primary items in the diet of kit foxes at Camp Roberts in northern San Luis Obispo County (Logan et al. 1992), western Merced County near the base of the Coast Ranges (Briden et al. 1992), and at the Bena Landfill northeast of Bakersfield (Cypher and Brown 2006). All these sites encompass grassland habitats on the margin of the San Joaquin kit fox range.

Smaller rodents, particularly pocket mice and deer mice are generally supplemental items that are likely captured opportunistically. They generally make up less than 10% of the diet; however, when kangaroo rat abundance is low, pocket mice in particular can make up a significant proportion of the diet or even be the primary food item. This commonly occurs during and immediately following extended droughts. During such events, kangaroo rats can decline to extremely low numbers. However, possibly because they can get by on smaller quantities of food or because they are not being suppressed by the more dominant kangaroo rats, pocket mice are usually still available during droughts. Pocket mice were the primary food for kit foxes during droughts on the Carrizo Plain in the 1990s (White et al. 1995) and the 2010s (Cypher, Fiehler et al. 2014; Cypher, Westall, et al. 2022).

Invertebrates play an interesting role in the foraging ecology of kit foxes. Invertebrates commonly consumed by kit foxes include Jerusalem crickets, other crickets, grasshoppers, a variety of beetles, beetle larvae, earwigs, and scorpions. It is not surprising that kit foxes consume these items. Invertebrates are highly nutritious. Also, they likely have little foraging cost in that they are probably consumed opportunistically while kit foxes are hunting rodents or other prey, and being small they are easy to capture and consume. What is a bit more surprising is that at times, invertebrates can constitute the primary food items in kit fox diets. As with other alternate prey, this high consumption of invertebrates is observed when abundance of kangaroo and also other vertebrate prey is low. On two different study sites on the Carrizo Plain, invertebrates were the dominant items consumed by kit foxes during 2014–2015 when rodent abundance, particularly of kangaroo rats, was extremely low due to a multi-year drought (Cypher, Fiehler, et al. 2014, Cypher, Westall, et al. 2022). On the study site on the northern end of the Carrizo Plain, no successful kit fox reproduction was recorded during this time; however, weights of foxes were typically at or above normal.

Other items that occur, usually at low frequencies, in the diet of kit foxes include lizards, snakes, and birds. As with other alternate items, these prey are likely encountered and consumed opportunistically while kit foxes are hunting for kangaroo rats. Also, lizards and birds are active primarily diurnally, which reduces the probability of kit foxes encountering them.

Seasonal variation in diet is not pronounced. Kit foxes tend to consume the same food items year-round. Kit fox diets rarely include items with strong seasonal variation in availability, such as fleshy fruits that seasonally can compose a large proportion of gray fox and red fox diets (Cypher 2003). Most of the primary dietary items tend to be available year-round. One exception may be the consumption of fewer invertebrates in the winter, when these items tend to be significantly less abundant. Reptiles also tend to be less abundant in winter.

Some spatial variation is observed in kit fox diets, but it is generally not pronounced and is largely predictable. The specific species of kangaroo rat consumed varies among locations based on which species occur in a given area. All things being equal, kit foxes usually prey more on the largest species that occurs in an area. Thus, giant kangaroo rats (131–180 g; Jameson and Peeters 1988) are consumed over Heermann's kangaroo rats (56–74 g; Jameson and Peeters 1988), and both are preferentially consumed over any of the subspecies of San Joaquin kangaroo rats (35–44 g; USFWS 1998). California ground squirrels, gophers, and invertebrates are commonly the primary items consumed by kit foxes near the margins of the range, where vegetation typically is higher and kangaroo rat abundance is lower.

Although not extensively investigated, diet likely does not vary significantly with age. Weaned pups still being provisioned at natal dens feed on whatever the adult foxes bring to the den. This is usually larger prey like rodents, rabbits, and occasional snakes and birds. When providing for pups, it is most efficient for the adults to bring larger items to the den to maximize the amount of food per trip and reduce the number of trips, which also reduces predation risk as well as leaves more time for other parenting activities (see Social Ecology). When the pups begin to attempt to forage on their own, they likely consume a higher proportion of invertebrates as these are generally easier to capture. The proportion of vertebrates in their diet likely increases as they gain experience in hunting.

SPACE USE, MOVEMENTS, DISPERSAL, AND ACTIVITY

In addition to food, water, and cover, space is also a critical resource for any species. An important attribute of the life history of a species is how it uses space such as establishing a home range, use of habitats within that home range, movements including dispersal, and active versus resting periods.

Home Range

Like virtually all terrestrial mammalian species, each kit fox occupies a particular area where it is familiar with the locations of critical resources (e.g., food, dens) and in which it fulfills all its needs. This area is commonly referred to as a "home range". In some species, such as wolves, this area is more often referred to as a "territory", but this term is generally used to refer to areas that are actively and aggressively defended from conspecifics (Burt 1943). There is little evidence to indicate that kit foxes engage in such active defense. Thus, home range is probably the more appropriate term.

For species that do not vigorously defend an area, the size of the home range is determined by the amount of area needed to fulfill life history requirements on a routine basis. Thus, home range size can vary with habitat quality and associated resource density. Generally, home range size among foxes is inversely related to food availability (Macdonald 1981; Fuller and Sievert 2001; Macdonald et al. 2004). This is particularly evident in comparing the size of kit fox home ranges in core population areas (see Distribution and Habitat Preferences), where habitat quality is high, with areas on the margin of the distributional range, where habitat quality is usually lower. Home range sizes reported in the literature vary considerably based on where and when a given study was conducted. Size also varies with the method used to calculate size. The 95% minimum convex polygon (MCP) method is commonly used to calculate home range size, and all the values presented here are 95% MCPs.

With those caveats in mind, in good quality habitat in a normal year with regard to food availability (i.e., food is neither scarce nor super-abundant), each pair of kit foxes generally uses a home range approximately 5.4 km^2 (2.1 mi^2) in size (Cypher et al. 2013). However, as has been emphasized, conditions in the San Joaquin Desert vary considerably among years. Consequently, home range estimates for San Joaquin kit foxes reported in different studies also vary considerably depending on the years the studies were conducted and the environmental conditions in those years (Table 4). In general, the estimates are smaller for study sites in core population areas and larger for those in satellite population areas.

Even over relatively short distances, home range size can vary with habitat quality and food availability. Home range size was assessed on four sites in the northern Carrizo ecosystem during 2014–2017 (Cypher, Boroski, et al. 2021). On the Topaz Solar Farms site, where food availability was relatively low due to previous habitat disturbance, mean home range size was 9.4 km^2 (3.6 mi^2) but was just 5.2 km^2 (2.0 mi^2) on an adjacent reference site where much less habitat disturbance had occurred. On the California Valley Solar Ranch just 5 km (3.1 mi) away from the Topaz sites, mean home range size was just 3.9 km^2 (1.5 mi^2) and was 4.2 km^2 (1.6 mi^2) on an associated reference site just 2 km (1.2 mi) farther away. Kangaroo rat abundance was considerably higher on the California Valley Solar Ranch sites relative to the Topaz Solar Farms sites.

Home range size can also vary temporally with changes in re-source availability. Three estimates are available from the southern and central part of the Carrizo Plain, where habitat quality is high and a favored food, giant kangaroo rats, is usually abundant. During a period of extended drought when food availability was extremely low, mean home range size was 7.6 km^2 (2.9 mi^2) (White and Ralls 1993). Mean size was 4.2 km^2 (1.6 mi^2) in 2015–2016 during a shorter dry period when prey availability was also low. In contrast, mean size was just 1.3 km^2 (0.5 mi^2) in 2017–2018, when multiple years of favorable conditions produced a super high abundance of giant kangaroo rats (Cypher, Westall, et al. 2022). In another example of temporal variation, home range size decreased significantly across three years on the Topaz Solar Farms, California Valley Solar Ranch, and associated reference sites as prey abundance increased each year as a result of higher precipitation (Cypher, Westall, et al. 2019).

Another interesting example of the effects of resources on space use occurs in the urban kit fox population inhabiting the city of Bakersfield. In this urban environment, food abundance is extremely high (see Kit Foxes and Urban Areas). Consequently, kit foxes can easily fulfill needs in a smaller space, and home ranges are extremely small, with a mean of just 0.78 km^2 (Cypher, Deatherage, et al. 2023).

All the examples above attest to the considerable ecological plasticity of San Joaquin kit foxes and their ability to adapt to varying environmental conditions. Such plasticity tends to be common among canids, but it is clearly evident in the home range variation among San Joaquin kit foxes.

Kit foxes are not strongly territorial, and home ranges of adjacent family groups commonly overlap to some degree. Home range overlap was reported to be about 50% in the Lokern area (Spiegel and Bradbury 1992) and about 35% in the Elk Hills area (Zoellick et al. 2002). Overlap also appears to vary with resource and kit fox density. During drought conditions on the Carrizo Plain, home range overlap between family groups was estimated to be 14% (White and Ralls 1993). Similarly, home ranges on the northern Carrizo Plain exhibited little overlap during a period of low food availability (Cypher, Fiehler, et al. 2014) except for those of family members (Figure 2.22). Conversely, home range overlap was extensive in the Semitropic Ridge area in northern Kern County (Figure 2.22), where food

TABLE 4. *Mean home range and core area size for adult San Joaquin kit foxes from locations throughout the range*

Location	Year(s)	No. foxes	Mean home range size (km2)						Source	Notes
			100% home range method[a]	100% home range size	95% home range method[a]	95% home range size	Core area method[a]	Core area size		
Bakersfield, Kern Co.	2001–04	45	100% MCP	0.78	–	–	–	–	Cypher, Deatherage, et al. 2023	Satellite population area; urban
Bakersfield, Kern Co.	1997	28	100% MCP	1.7	95% FK	1.2[b]	50% FK	0.2	Frost 2005	Satellite population area; urban
Carrizo Plain, San Luis Obispo Co.	2017–18	12	100% MCP	2.3	95% MCP	1.3[b]	50% MCP	0.4	Cypher, Westall, et al. 2022	Core population area; natural lands; high prey availability
Tupman–Buttonwillow, western Kern Co.	1977	9	100% MCP	2.5	(Est 95% MCP)	1.7[b]	–	–	Knapp 1978	Core population area but lands actively being converted to agriculture
Panoche Valley, eastern San Benito Co.	2019–22	43	–	–	95% MCP	2.4	50% MCP	0.4	Cypher, Westall, et al. 2023	Core population area; natural lands; reference site for Panoche Valley Solar Farm study Year 1; high prey availability
Semitropic Ecological Reserve, northern Kern Co.	2011–12	7	100% MCP	3.7	95% MCP	2.4[b]	50% MCP	0.9	Cypher, Westall, et al. 2014	Satellite population area; natural lands
Elk Hills, western Kern Co.	1994	8	100% MCP	4.3	(Est 95% MCP)	2.8[b]	75% HM	1.3	Cypher, Koopman, et al. 2001	Core population area; natural lands (only June–Sept)
Elk Hills, western Kern Co.	1984–85	9	100% MCP	4.3	(Est 95% MCP)	2.8	50% HM	1.2	Zoellick et al. 2002	Core population area; natural lands; includes juveniles
Elk Hills, western Kern Co.	1980–86	12	100% MCP	4.8	(Est 95% MCP)	3.1	50% HM	1.2	Zoellick et al. 2002	Core population area; oil and gas production; includes juveniles
California Valley Solar Ranch, San Luis Obispo Co.	2014–17	23	–	–	95% MCP	3.9	50% MCP	0.9	HTH 2019	Core population area; natural lands; solar site
California Valley Solar Ranch, San Luis Obispo Co.	2014–17	22	–	–	95% MCP	4.2	50% MCP	1.1	HTH 2019	Core population area; natural lands; reference site for California Valley Solar Ranch study

Mean home range size (km2)

Location	Year(s)	No. foxes	100% home range method[a]	100% home range size	95% home range method[a]	95% home range size	Core area method[a]	Core area size	Source	Notes
Carrizo Plain, San Luis Obispo Co.	2015–16	22	–	–	95% MCP	4.3	50% MCP	1.1	Cypher, Westall, et al. 2022	Core population area; natural lands; low prey availability
Lokern Natural Area, western Kern Co.	1989–93	12	–	–	95% MCP	4.8	50% HM	1	Spiegel 1996	Core population area; oil and gas production; includes juveniles
Topaz Solar Farms, San Luis Obispo Co.	2014–17	32	–	–	95% MCP	5.1	50% MCP	1.2	Cypher, Westall, et al. 2019	Core population area; natural lands; reference site for Topaz Solar Farm study
Lokern Natural Area, western Kern Co.	2001–04	32	–	–	95% FK	5.9	–	–	Nelson et al. 2007	Core population area; natural lands
Panoche Valley, eastern San Benito Co.	2019–22	32	–	–	95% MCP	6.1	50% MCP	1.5	Cypher, Westall, et al. 2023	Core population area; Solar site Year 1; high prey availability
Northern Carrizo Plain, San Luis Obispo Co.	2012–13	9	100% MCP	9.9	95% MCP	6.3	50% MCP	2.0	Cypher, Fiehler, et al. 2014	Core population area; natural lands
Lokern Natural Area, western Kern Co.	1989–93	14	–	–	95% MCP	6.7	50% HM	1.4	Spiegel 1996	Core population area; natural lands; includes juveniles
Carrizo Plain, eastern San Luis Obispo Co.	1989–91	21	100% MCP	11.6	(Est 95% MCP)	7.6	–	–	White and Ralls 1993	Core population area; natural lands; low prey availability due to drought
Topaz Solar Farms, San Luis Obispo Co.	2014–17	19	–	–	95% MCP	9.4	50% MCP	2	Cypher, Westall, et al. 2019	Core population area; solar site
Panoche Valley, eastern San Benito Co.	2015–16	6	–	–	95% MCP	11.4[b]	50% MCP	1.8	Cypher et al. unpubl. data	Core population area; natural lands; pre-solar farm construction; low prey availability

[a]MCP = minimum convex polygon; FK = fixed kernel; HM = harmonic mean; Est 95% = 100 MCP/1.53

[b]One-year study.

Note: Data are ordered from lowest to highest 95% home range size.

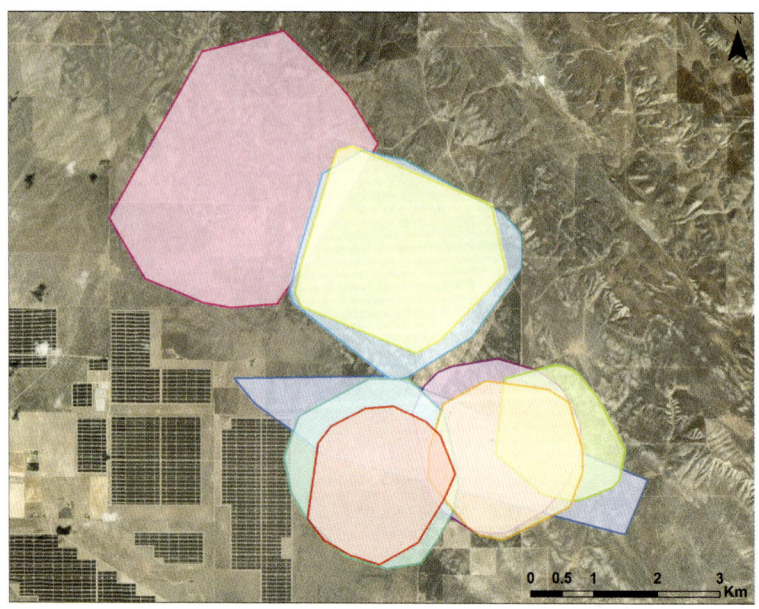

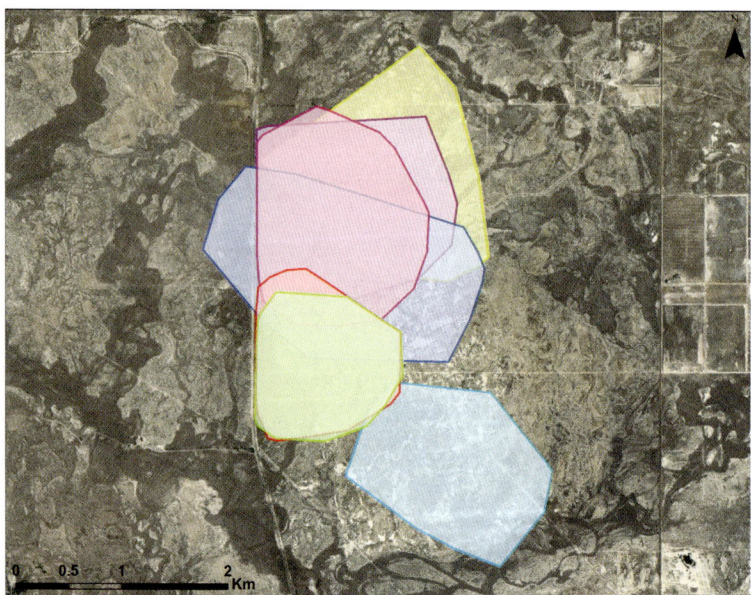

FIGURE 2.22. San Joaquin kit fox home range overlap (a) on the Northern Carrizo Ecological Reserve where food availability was relatively low, and (b) on the Northern Semitropic Ridge Ecological Reserve where food was abundant and concentrated. Images prepared by Tory Westall.

availability was high and suitable habitat was limited, resulting in multiple foxes using a relatively small area (Cypher, Westall, et al. 2014). Considerable overlap was also observed among foxes in Bakersfield where resource abundance and fox density were high (Cypher unpublished data).

Exactly how kit foxes space themselves intraspecifically is not clearly known. Unlike larger canids such as wolves and coyotes, there is little to no evidence that kit foxes actually defend home ranges through aggressive interactions. Kit foxes have not been found injured or dead with evidence pointing to another kit fox as the cause. If such intraspecific aggression does occur, it may be limited to a resident fox simply chasing an intruding individual out of its home range. Spacing may be achieved more through scent-marking. Kit foxes extensively scent-mark (Murdoch 2004; Ralls and Smith 2004; Clark 2007), and such marking increases in frequency around the margins of a home range (Murdoch 2004). Such passive defense of a home range may help limit competition for resources as well as reduce the considerable risk of injury and energy expenditure associated with aggressive defense of an area.

Home range core areas are areas of intensive use within home ranges. Core areas are commonly defined by calculating a 50% MCP. Although home ranges may overlap, core areas of foxes from different family groups almost never overlap. These areas may actually be defended from any intruding kit foxes, particularly if the core area encompasses a natal den.

Habitat Use

Habitat use by kit foxes is not an extensive topic as it is for many other species, primarily because most of the areas used by kit foxes tend to be rather homogeneous in habitat composition and commonly comprise just one basic type, usually arid scrub or grassland (see Distribution and Habitat Preferences). Use of anthropogenic habitats is discussed in Chapter 3. Use of other natural habitat types typically occurs only along the edges of kit fox habitat or in the marginal areas of the range. In the marginal areas of the range, kit foxes will sometimes use oak woodland savanna if the density of trees is not too great. Use of this type was documented in the Salinas Valley on Camp Roberts and Fort Hunter-Liggett (O'Farrell et al. 1987; JSA 1995).

Some relatively common habitat types in the San Joaquin Valley are largely avoided by kit foxes and warrant emphasis. Wetland and riparian areas are typically avoided. These habitat types used to be much more extensive in the San Joaquin Valley (USFWS 1998; Kelly et al. 2005) but are still present in some abundance. In these areas, the dense vegetation excludes favored kit fox prey, inhibits kit fox movements, and conceals kit fox predators. Agricultural lands are also generally avoided (see Kit Foxes and Agricultural Lands). Common practices such as frequent disking, use of biocides, rodent control, and flood irrigation all result in a lack of food and dens for kit foxes. Also, coyotes, dogs, and red foxes all commonly occur in agricultural areas. One exception is orchards (e.g., nut, citrus), particularly those watered via drip irrigation. Limited prey are sometimes available, but even these habitats are used only infrequently by kit foxes and only when they abut natural lands.

Movements

Routine kit fox movements are not especially remarkable. Most activity begins around whatever den a fox uses during that particular day. During the pup-rearing season, activity is concentrated around natal dens, with the adults making forays from the dens however far necessary to secure food for themselves or their pups. When pups are not present, foxes will venture out to forage and may eventually return to the same den or end the evening at a different den where they then spend the next day. Zoellick et al. (2002) estimated that San Joaquin kit foxes travel an average of 8–16 km (5–10 mi) per night. The longer distances were associated with the breeding and pup-rearing season, when foxes were seeking mates or extra-pair copulations or searching for food to provision pups.

In a study conducted on urban kit foxes, individuals with radio collars were intensively followed on foot after they emerged from their dens in the evening. Many of these foxes, but particularly adult males, traveled essentially around the entirety of their home ranges in the course of one night and extensively scent-marked (Murdoch

2004). Anecdotal information suggests that foxes in natural habitats do the same.

Typical of canids, kit foxes occasionally make long-distance movements that do not seem to constitute dispersal. These may best be described as exploratory forays. Most typically, they are observed among adult foxes. An adult male with an established home range in the southern Carrizo Plain was documented traveling ~30 km (18 mi) to the northern part of the Plain, and then returning to his home range. An adult female on the southern Panoche Valley made repeated trips of about 15 km (9 mi) over rugged terrain to a smaller valley to the south, where she would spend one to several days and then return to the Panoche Valley (Cypher, Westfall, et al. 2023). These are two examples of extreme movements. Other foxes have been documented making similar but shorter movements outside their home ranges. All these foxes were adults with established home ranges in areas where food seemed plentiful, and they always returned to their home range. Thus, the movements were not dispersals, and the foxes did not seem to be seeking areas with more food. A possible explanation is that they may have lost mates and were seeking new ones.

Some foxes also engage in shorter movements into neighboring home ranges and then quickly return to their own range. This behavior is observed primarily in males, particularly around the breeding season. These males may be checking on the estrous condition of neighboring females and then engaging in extra-pair copulations (see Social Ecology).

Only recently have GPS-enabled collars become light enough in weight to place on kit foxes. These collars can collect a wealth of detailed information, including more comprehensive and detailed data on kit fox movements. The long-distance movements described above may be more common than previously realized. They are unlikely to be detected using conventional VHF collars. Thus, more detailed information on kit fox movements will likely come to light as additional studies are conducted.

Dispersal

Dispersal movements by kit foxes do occur, and these movements are typically exhibited by foxes less than 2 years old. Some foxes disperse in their first year, while others delay dispersal until their second year or sometimes even later (see Social Ecology). Not uncommonly, dispersing foxes will make one or more exploratory movements to new areas and then return to their natal area prior to permanently dispersing. Others just leave and never return. Routine dispersal movements are notoriously difficult to quantify, as not uncommonly the animals move out of the study area and are then hard to locate. These animals are sometimes detected via aerial telemetry searches, opportunistic captures on another study site, or when dispersing animals are struck by vehicles while crossing roads. The most detailed dispersal information came out of a 15-year study of San Joaquin kit foxes at the Naval Petroleum Reserves in Kern County. The study area was large (314 km^2 or 121 mi^2), and researchers searched for kit foxes over the entire area and also conducted occasional aerial searches. Thus, there was a greater probability of finding foxes that made routine dispersal movements. These data were summarized in Koopman et al. (2000). They found that of 209 juvenile kit foxes monitored during 1980–1996, 33% dispersed from their natal territory. Significantly more males (49.4%) than females (23.8%) dispersed. The age range of dispersing foxes was 4–32 months, although dispersal peaked in July-August of the birth year. Mean dispersal distance was 7.8 km or 4.8 mi (range 1.8–32.5 km or 1.1–20.2 mi) and was similar for males and females (Scrivner et al. 1987).

Finally, some foxes occasionally engage in extreme dispersal movements. A fox originally marked at Camp Roberts was found dead on a road in the Carrizo Plain, a distance of 95 km (59 mi). Another marked fox from Elk Hills was found dead on a road in the Cholame Pass area, a distance of approximately 124 km (77 mi) (Scrivner et al. 2016). Yet another fox marked at Elk Hills was detected by aerial telemetry over on the Carrizo Plain, a distance of 25 km (15.5 mi) over a mountain range. Finally, in what was maybe more of a homing movement than a dispersal, a fox translocated from Bakersfield to Elk Hills left its release site and traveled 24 km (14.9 mi) back toward Bakersfield before being struck by a vehicle (Scrivner et al. 2016). All these movements indicate that kit foxes, although a relatively small in size, have the capacity to move extraordinary distances at times.

Activity

Kit foxes are primarily nocturnal, but otherwise not much detailed information is available on activity patterns of kit foxes. They typically emerge from their dens at dusk and reenter the dens around dawn. During the night hours, it is assumed they are foraging, exploring, interacting, and engaging in other activities that fulfill life history needs. Based on observations during the intensive-follow studies with urban kit foxes, kit foxes commonly emerged from dens, foraged, quickly toured the perimeter of their entire home range, and then after 3 or 4 hours, lay down out in the open and slept for one or more hours. They then became active again an hour or two prior to dawn, at which time they reentered dens.

There are a few exceptions to the general pattern above. Kit foxes have been seen out traveling during the day. Although not proven, these may be animals in the process of dispersing. They may not be familiar with den locations in the unfamiliar areas they are moving through resulting in more aboveground activity. Foxes residing in areas where California ground squirrels make up the primary prey may have to forage before sunset or after sunrise so that their hunting coincides with the activity patterns of the mostly diurnal squirrels. A few anecdotal observations support this. A final exception is during pup-rearing. As the pups get older, their activity levels increase, and they not uncommonly come aboveground during the day to play. When they are aboveground, it is not uncommon for one or more adults to also be outside the den to keep an eye on the pups and watch for predators.

DENS

Kit foxes are obligate den users. Kit foxes and some other arid land foxes have evolved such that den use is essential to their survival. Thus, kit foxes and closely related swift foxes are unique among North America canids in that they use a den every single day of the year. All other canids in North America, if they use a den at all, just use it during reproduction to bear and rear young. Dens are used by kit foxes for daytime resting, avoiding extreme temperatures or weather, conserving moisture, bearing and rearing young, and eluding predators (Grinnell et al. 1937; Koopman et al. 1998). Thus, den use is a critical and integral aspect of their ecology.

Nocturnal activity and diurnal den use are behavioral adaptations that facilitate living in environments that are arid and where warm season temperatures can be quite extreme. Thus, avoiding direct exposure to the sun and spending the day in a den where the relative humidity is higher than ambient conditions outside the den help kit foxes avoid overheating and reduce moisture loss. Loredo et al. (2020) found that temperatures just 2 m (6.6 ft) inside of dens can be 6.5% cooler than ambient temperature on average in summer (28.9 °C or 84.0 °F in the den versus 30.9 °C or 87.6 °F ambient) and that relative humidity can be 40.5% higher (69.4% in the den versus 49.4% ambient). Based on measurements collected by Cypher, Murdoch, et al. (2021), temperatures in dens can be as much as 31.1% cooler in summer and 479.3% more humid! Thus, dens significantly facilitate thermoregulation and moisture conservation by kit foxes.

Dens also provide escape cover for foxes from predators such as coyotes, bobcats, dogs, and golden eagles. In a sense, dens are to kit foxes what trees are to gray foxes. Tree-climbing likely facilitates the survival of gray foxes in areas where coyotes are present (Cypher 1993, 2003). Similarly, den use may be a primary reason that kit foxes can co-occur with coyotes.

Kit foxes are strong diggers and can excavate their own dens; however, very commonly, they are inclined to conserve energy by not starting from scratch. They will usurp and modify burrows of other species, not uncommonly their prey species, such as kangaroo rats, gophers, and ground squirrels. They will also enlarge badger digs (holes dug by badgers in pursuit of prey). Another strategy is to find a spot where erosion or some other agent has created openings in the ground. Taking advantage of any of these situations saves considerable work and energy. Although kit foxes will readily use the burrows and dens of other species, they consistently avoid dens, even abandoned ones, created or used by larger species such as coyotes and red foxes. These dens are clearly large enough for these predators to enter and therefore not safe refugia for kit foxes.

As to locations chosen by kit foxes to construct their dens, there are no firm generalizations. Dens are found in a range of soils based on the predominant type in a given area occupied by kit foxes; however, when choices are available, kit foxes usually prefer well-drained soils such as sandy loams (Reese et al. 1992). Local topography sometimes appears to influence den placement. Kit fox dens are commonly found on gentle slopes if they are available in a given location. The slopes allow the foxes to dig more horizontally instead of vertically, and therefore they are working less against gravity when constructing a den. Slopes also provide kit foxes with a bit of a vantage point from which they can scan the surrounding nearby area for predators before moving too far from the protective cover of the den. If any areas are avoided for den construction, it would be low-lying areas or areas of water flow (e.g., wash or arroyo bottoms) where the dens might be flooded during precipitation events.

Den Characteristics

Although some attempts have been made to characterize kit fox dens, these characterizations are general at best, as the dens tend to be quite variable. This variability is attributable to factors such as age of the den, origin (e.g., created by a fox or initially belonging to another species), soil type, and topographic conditions. Loose sandy soils are softer and more erodible, and therefore the edges of den entrances collapse on their own or break with use by the foxes over time. Thus, dens in these soils can have larger entrances with great variability in shape; however, in loam or clay soils that are more compact, den entrances commonly have a keyhole shape to them (Figure 2.23). This shape, wider at the top and narrower at the bottom, nicely accommodates the wider body and narrower legs of kit foxes as they enter and exit. The entrance tends to be just a bit larger than the body of an adult kit fox, and this excludes entry by larger predators. Indeed, even if a den entrance does not have this keyhole shape, the tunnel beyond the entrance quickly narrows down such that it is difficult for any animal larger than a kit fox to enter. Dens initially dug by foxes or that began as a smaller burrow, such as that of a kangaroo rat, also commonly have this shape. Dens that were enlarged from ground squirrel burrows or badger digs commonly

FIGURE 2.23. "Keyhole"-shaped entrance typical of many San Joaquin kit fox dens. Photo by Erica Kelly.

have more of a circular shape to the entrances (Reese et al. 1992). Den characteristics have been quantified in several studies (Table 5). Although most dens have larger entrances, foxes have been observed using dens with entrances only about 8 cm in diameter. This ability to squeeze into small openings may serve them well when pursued by a larger predator.

TABLE 5. *Dimensions and number of entrances for San Joaquin kit fox dens*

Location	No. dens	Entrance height (cm) Mean	Range	Entrance width (cm) Mean	Range	Mean no. entrances	Range	Source
Elk Hills, Kern Co.	882	19.1	8.9–68.1	18.0	19.1–51.1	3.4	1–17	Berry et al. 1987
Lokern, Kern Co.	304	28.2	9.5–85.0	21.3	7.0–75.0	2.6	1–11	Spiegel 1996
Camp Roberts, San Luis Obispo and Monterey Co.	717	20.0	9.0–40.0	21.0	10.0–43.0	–	–	Reese et al. 1992

FIGURE 2.24. Long "berms" of excavated material are common outside of kit fox dens. Photo by Ellen Cypher.

Little is known about the internal structure of dens, although again, variability is likely the norm. The structure depends on the substrate, the age of the den, and whether it is a natal or non-natal den. Morrell (1972) carefully excavated and mapped the internal structures of a couple of dens. The largest had nine entrances and at least six chambers. The internal runways or tunnels were generally 8–15 cm (3–6 in) high and 15–20 cm (6–8 in) wide. The chambers typically were 25 cm (10 in) high. The deepest location in the den was approximately 1.3 m (4.3 ft) below the ground surface.

A common characteristic observed at kit fox dens is the presence of a long berm or trail of dirt extending out from a den entrance. When most other species dig a den or burrow, the excavated material is commonly ejected in all or at least multiple directions. This results in more of a circular pattern of ejecta as is commonly seen around ground squirrel and badger burrows. Kit foxes, on the other hand, have an interesting habit of pushing all ejecta in just one direction creating a long berm (Figure 2.24). Some berms can be over 5 m (16 ft) long, and they may be straight or sinuous. It is unclear whether structuring the ejecta this way serves some particular purpose or is just a behavioral quirk. However, it does help biologists conducting surveys for kit foxes to distinguish their dens from those of other species!

Natal dens are a bit of a special case. These are dens where pups are reared. These dens are commonly traditional, meaning that a pair of foxes or whoever next inherits the home range (typically individuals related to one or both parents—see Social Ecology) will use the same den each year to raise the families. Partly, this may be a function of the success of the site. Because the successful production of young is a large component of fitness, it makes sense to return to natal dens where litters have been successfully reared. Also, natal dens need to accommodate a relatively large number of individuals. Litters as large as nine pups have been documented (see Reproduction). If both parents are also using the den and if any helpers are present, then there are that many more bodies needing space. Thus, it makes sense to return to a den that is already relatively large. As with other dens, natal dens are commonly modified and enlarged with time. This process is facilitated by the presence of many individuals that can help dig. Even the pups begin digging at a relatively early age and help with enlarging the den.

Because of all the material brought to the surface during the excavating, older natal dens sometimes have a somewhat mounded appearance. Also, because of all the excavating, natal dens almost always have multiple entrances (Figure 2.25). This is not only a function of modification over time and, but also a matter of practicality, as many bodies are constantly entering and exiting the den during active periods. The ability for many bodies to get into the den extremely rapidly is critical if a predator approaches the den trying to pick off a pup.

In addition to the characteristics above (e.g., multiple entrances, lots of digging), natal dens can also be identified by an abundance of prey remains and pup scats (Figure 2.26). Once weaning begins, the parents and helpers continually bring back prey for the pups. Food items may not be consumed in their entirety. Thus, it is not uncommon to find tails, heads, feet, or pieces of skin of kangaroo rats, rabbits, squirrels, snakes, and other prey lying around the den area. Particularly in years of prey abundance or if a litter is relatively

FIGURE 2.25. Natal den of a San Joaquin kit fox with multiple entrances. Photo by Brian Cypher.

small, entire unconsumed carcasses of prey may even be present at the den. Not surprisingly because of the concentration of activity around the dens, they are commonly littered with numerous small pup scats, which serves as another distinguishing feature.

Some dens used by kit foxes are termed "atypical". These are dens that are usually located in man-made structures. Atypical dens include culverts, pipes (in-ground and on the surface), rock and wood piles, beneath concrete slabs or asphalt surfaces, and under seatrains and buildings. Atypical dens are more commonly used in altered environments such as oil fields (Berry et al. 1987; Spiegel 1996), military installations (Reese et al. 1992), and urban areas (Cypher 2010). This adaptability and propensity to use man-made structures has led to a successful conservation strategy for San Joaquin kit foxes: the use of artificial dens to replace or supplement natural dens. A few artificial dens were first installed in the 1980s and 1990s, either to replace natural dens that had been destroyed during development projects or to provide dens for foxes in locations where dens were not naturally abundant. The dens were made from a variety of materials, includ-

FIGURE 2.26. (a) Pup scats and (b) kangaroo rat remains outside of San Joaquin kit fox natal dens. Photos by Brian Cypher.

A

B

ing polyvinyl chloride (PVC) pipe, metal pipe, or high-density poly-ethylene (HDPE) plastic pipe. Some dens were placed on the surface of the ground and covered with dirt while others were installed underground with some sort of chamber. The dens were not regularly monitored, although some use by foxes was noted.

During 2001–2004, a study was conducted to determine whether kit foxes preferred particular den designs or materials (Cypher, Murdoch, et al. 2021). The designs tested included surface dens of different lengths and subterranean dens with and without chambers and with one or two entrances (Figure 2.27). The dens were constructed of concrete, aluminum, PVC, and HDPE. The two types of chambers tested were an irrigation valve box and a small "dogloo" doghouse. A total of 34 dens were installed and monitored weekly to assess use by kit foxes. The foxes used all the designs and materials, and no strong preferences were detected (Figure 2.28).

Recommendations based on these results included installing surface dens about 6 m (20 ft) long in areas where additional rapid escape cover might benefit kit foxes. In a study in Texas, the installation of these surface dens improved the survival of swift foxes (McGee et al. 2006). In areas where additional subterranean dens are desired for kit foxes, either to replace natural dens that have been destroyed or to enhance den availability, Cypher, Murdoch, et al. (2021) recommended a design that was easy to construct with materials costing about $100 (Figure 2.29). This design consists of two flexible single-wall HDPE pipes that are 2–3 m (6–10 ft) in length and that connect to an underground irrigation valve box in which holes have been cut in the sides.

Den Use Patterns

Each kit fox uses multiple dens during the course of a year. The number varies but the average is generally between 10 and 20 dens per year (Table 6); however, estimates for individual foxes have been as high as 64 (Reese et al. 1992). There is no predictable pattern in the number of consecutive days a fox will use a given den before switching to another den. Presumably, if prey is abundant in a given area and no particular threats are present, a fox might use a given den for multiple days. Even natal dens are occasionally switched. While

A

FIGURE 2.27. Some of the artificial den designs and materials tested to determine kit fox preferences: (a) an HDPE surface den, (b) a one-entrance concrete den with an irrigation valve box chamber, and (c) a two-entrance concrete den without a chamber installed above a two-entrance aluminum den with a dogloo chamber. Photos by Brian Cypher.

B

C

A

B

FIGURE 2.28. Kit foxes using (a) a concrete surface den and (b) a subterranean HDPE den. Photos by (a) Brian Cypher and (b) Francesca J. Ferrara.

FIGURE 2.29. Recommended artificial den design and materials for San Joaquin kit foxes: a two-entrance den with HDPE tubes for tunnels and an irrigation valve box chamber. Photo by Brian Cypher.

the pups are still young (e.g., ~4 weeks or less), a family group may remain in the same den; however, most families eventually move to new dens. These moves may be a strategy to escape high flea populations that build up in dens (see Survival and Mortality Factors). The speed at which flea populations increase in the new den may depend on factors such as the size of the family group, soil moisture, and temperatures.

There also seems to be no particular pattern to the distribution of dens within a home range. Dens are generally dispersed throughout the home range, which is advantageous for eluding predators that might be encountered while foraging or engaging in other movements around the range. Natal dens tend to be located more toward the interior of a home range, possibly to reduce disturbance or interference from neighbors. Dens can also be clumped, particularly dens used to rear pups. This may be a function of not wanting to move the pups too far (longer distances result in greater exposure to predation) while also wanting to move to escape fleas, as discussed above.

TABLE 6. *Den use by San Joaquin kit foxes from locations throughout the range*

Location	Year(s)	No. foxes	Mean dens used per fox	Range	Source	Notes
Topaz Solar Farms, San Luis Obispo Co.	2014–17	45	8.4	1–31	Cypher, Westall, et al. 2019	Core population area; natural lands; reference site for Topaz Solar Farm study
Topaz Solar Farms, San Luis Obispo Co.	2014–17	26	11.2	1–33	Cypher, Westall, et al. 2019	Core population area; solar site
Elk Hills, western Kern Co.	1980–87	20	11.4	–	Koopman et al. 1998	Core population area; oil developed
Elk Hills, western Kern Co.	1980–87	26	12.6	–	Koopman et al. 1998	Core population area; natural lands
Lokern, western Kern Co.	1989–93	14	13.6	5–21	Spiegel unpubl. data	Core population area; oil developed
California Valley Solar Ranch, San Luis Obispo Co.	2014–17	13	15.1	–	HTH 2019	Core population area; solar site
Lokern, western Kern Co.	1989–93	25	15.6	9–25	Spiegel unpubl. data	Core population area; natural lands
Western Merced Co.	1985–87	28	16.0	1–58	Briden et al. 1992	Satellite population area; natural lands
Camp Roberts, northern San Luis Obispo Co.	1989–91	86	17.6	1–64	Reese et al. 1992	Satellite population area; natural lands with occasional military training activities
California Valley Solar Ranch, San Luis Obispo Co.	2014–17	10	19.4	–	HTH 2019	Core population area; natural lands; reference site for California Valley Solar Ranch study

Note: Data are ordered from lowest to highest mean dens used per fox.

Not much is known about use of kit fox dens by other species. It would be surprising if the dens, particularly unoccupied ones, were not used by a diversity of vertebrates and invertebrates. Some interesting observations have been recorded. Burrowing owls have been observed using kit fox dens in what appears to be a dynamic situation. Foxes have been observed using a den, and then they apparently move on to another den. During this vacant period, burrowing owls have been observed using the den for some period (see Interspecific

Interactions). Later, foxes have been observed again using the den. It is not clear whether each species is just taking advantage of a vacant den or whether either is influencing use by the other through displacement. In urban environments, a similar dynamic has been observed not only among burrowing owls but also with non-native red foxes, striped skunks, California ground squirrels, and feral cats.

In the situations above where burrowing owls and other mammals use kit fox dens, the assumption is that two or more species (kit foxes and the others) do not use the dens simultaneously. However, a rather fascinating situation has been observed in urban environments of simultaneous use of a den by kit foxes and other species. Burrowing owls and California ground squirrels have been observed entering dens occupied by radio-collared kit foxes. In addition, radio-collared kit foxes and radio-collared striped skunks were tracked to a single den on at least four occasions in Bakersfield (Harrison et al. 2011). The circumstances of such situations, particularly whether any aggressive encounters occur in the dens, are not known.

SOCIAL ECOLOGY

The social system of kit foxes is consistent with that typical for small canids (Moehlman 1989). In brief, they exhibit monogamy with occasional polygamy, pairs typically mate for life, both parents participate in pup-rearing, helpers (typically young from the previous year) may be present and may assist with pup-rearing, family members commonly share dens, and young may delay dispersal for more than a year.

Like most canids, kit foxes are monogamous (Ralls et al. 2007). Monogamy is actually rare among mammals. A more precise and detailed characterization may be that kit foxes exhibit social monogamy that is also obligatory and perennial. Each of these terms warrant additional explanation. Obligatory monogamy occurs when reproductive success is significantly enhanced by both parents' participating in the rearing of young. This enhancement could be in the form of a higher proportion of pups surviving to dispersal age, or the young being in better condition because both parents are provisioning them. Kit fox monogamy is perennial because pair-bonds last multiple years and commonly last until one member of the pair dies.

Ralls et al. (2007) reported that one pair of kit foxes remained together for 701 days. They reported that this was actually a minimum as the pair was already together when monitoring first began, and the pair was still together when radio contact was eventually lost because of battery expenditure. In Bakersfield, where longevity tends to be higher among the urban foxes, two pairs were documented to have produced litters in three consecutive breeding seasons (Cypher unpublished data).

Although kit foxes are socially monogamous, they are not necessarily genetically monogamous. Socially monogamous indicates that a pair remains together and rears their young together. However, male kit foxes occasionally engage in polygyny through extra-pair copulations, and apparently females will readily breed with a non-mate male (Ralls et al. 2001, 2007; Murdoch et al. 2008b; Westall et al. 2018). This is discussed further below.

Pairs can form at any time of year; however, if adults are unpaired, there seems to be an extra effort to find a mate in the fall before the breeding season begins (typically in late November–early December). Although they are biologically able to reproduce in their first year, young foxes commonly do not pair until sometime in their second year and then breed during the next breeding season. Kit foxes suffer relatively high mortality rates, and therefore the chances are good that a fox may have at least two different mates during its lifetime.

Kit foxes occasionally engage in extra-pair copulations (Ralls et al. 2001, 2007; Murdoch et al. 2008b; Westall et al. 2018). Put more bluntly, they sometimes cheat on their mates! When this occurs, it typically consists of a male from one pair traveling into the neighboring home range and mating with the resident adult female. Females are only estrus for about four days, a relatively short period of time. How exactly neighboring males know when to trespass for the attempted copulation is unknown. Also, given the short period of estrus, it is somewhat surprising that the resident male apparently does not engage in mate guarding. Finally, the resident female appears to readily engage in the affair. Although not documented, there is no reason to believe that a given male might not engage in extra-pair copulations with more than one neighboring female. Nor is there any reason to believe that the mates of these males do not also copulate with one of the neighboring males. The result of this opportunistic promiscuity is that many kit fox litters have mixed paternity.

Clearly, there is a fitness benefit to males that engage in extra-pair copulations. They are able to sire more pups but do not have to raise them all. Females that engage in these affairs also potentially realize fitness benefits because with mixed paternity, their litters are more genetically diverse (Westall et al. 2018), and this may enhance genetic fitness and improve the odds that some pups will be successful. Thus, for the females, extra-pair copulations increase the opportunity to secure superior paternal genes for the pups, and it also decreases the chances of potential litter failure due to mating with a genetically incompatible male (Zeh and Zeh 2001).

Both members of a mated pair participate in the rearing of the pups. Parental care by the mother is not surprising and, in fact, is essential, at least until the pups are weaned (Westall et al. 2019). The mother is the sole source of nutrition for the pups for their first 3–4 weeks, and the pups may continue to nurse until about 8 weeks old. If something happens to the female before the pups are weaned, the pups will starve and perish. During this critical early period, the father assists by bringing prey items back to the natal den for the female. This allows her to conserve her energy for lactation, and not having to leave the den to forage also reduces her exposure to predators. As the pups become older, and particularly as they begin the weaning process, the mother begins to leave the den more frequently, in particular to help find prey to feed the growing pups.

Once the pups begin transitioning from milk to solid food, the father is able to raise the litter in the event that something happens to the mother. Potentially, even a "helper" fox could raise the litter alone if something happened to both parents, as has been documented on several occasions (Spiegel 1996; Koopman et al. 2000). The success rates of litters raised by a single parent or individual has not been quantified but is likely lower.

Foxes known as "helpers" are sometimes present at natal dens (Figure 2.30). These individuals are typically young from a previous year's litter that have chosen to delay dispersal and remain in their natal home range. The advantage to them is that they remain in a familiar area and gain additional experience in finding food and eluding predators. If they assist with rearing the pups, they may also gain

FIGURE 2.30. An adult female (front), adult male (right), and female helper (background) attending to pups at a natal den. Photo by Tory Westall.

experience in rearing prior to having their own litters. Presuming they are related to the pups, they may also be contributing to their own genetic fitness if their assistance enhances the success of the litter. In most if not all circumstances, helpers are related to one or both parents. If one of their parents subsequently dies and their remaining parent pairs with a new individual, the helpers would clearly be related to only one of the parents. No situation has been documented to date in which a helper was not related to one of the parents.

Helpers among San Joaquin kit foxes are typically female (Koopman et al. 2000; Ralls et al. 2001), but some number of cases of male helpers have been documented (Spencer et al. 1992; Westall et al. 2018). Also, typically there is only one helper present, but as many as four have been documented (Westall et al. 2018). The presence and number of helpers may be dependent on resource availability. The Resource Dispersion Hypothesis (Macdonald 1981) posits that additional individuals will be tolerated within a home range if resources are abundant. Consistent with this hypothesis, helpers tend to be much more commonly observed in urban environments where resources tend to be consistently super abundant (see Kit Foxes and Urban Areas). Indeed, the family group with four helpers occurred in an urban kit fox population. Interestingly, at least two instances have been documented of helpers that were "grandmothers" (Cypher unpublished data). In both cases, the older females had raised litters within a given home range area, and then suddenly one of their daughters became the reproducing female in that home range. It is not known whether the older females became barren due to age, whether there was some shift in dominance within the family group, or some other factor that resulted in the switch. Regardless, the older females were documented assisting with the rearing of the younger female's litter. Both of these occurrences again were in an urban fox population, where the life span of foxes is considerably longer. Possibly, foxes in natural habitats do not live sufficiently long for this role reversal to occur.

The role played by helpers in assisting with pup-rearing varies (Westall et al. 2018). Some helpers just seem to be present without providing much obvious assistance, although it could be that they are providing a subtle form of vigilance. Other helpers provide much more obvious assistance, such as playing with pups, being vigilant, and even bringing prey back to the den. The benefits of this assistance in terms of pup survival or fitness have not been quantified, but the fact that it occurs suggests evolutionarily that there likely is a beneficial effect. As mentioned, there are benefits for the helpers in that they are able to remain in familiar territory and gain experience, and they also sometimes inherit the home range if the parents perish before the younger fox disperses.

In rare instances, two family groups have been raised together (Figure 2.31). This has been documented on at least four occasions (Zoellick et al. 1987; Westall et al. 2019; ESRP unpublished data). In two cases, the females rearing their litters together were known to be related (mother and daughter; Westall et al. 2019). In these two cases, allonursing was also observed, in which both females nursed pups from both litters.

Adoption has also been documented among San Joaquin kit foxes, although in both cases the adoption was human-assisted (Cypher, Westall, et al. 2021; Deatherage et al. 2023). In 1992, a radio-collared

FIGURE 2.31. Two female kit foxes at Bakersfield College that have both reproduced and are raising litters of pups together. Both have enlarged teats from nursing. The female on the left has been dye-marked. Photo by Tory Westall.

female kit fox was found dead (killed by a coyote), and she was heavily lactating. The biologists excavated her natal den and recovered six pups approximately 3 weeks old. The pups were raised by a rehabilitator until about 10 weeks old. Four of the pups were placed back in the den of their father (also radio-collared). A new female was present, and the pair accepted and raised the pups. The new female was possibly a daughter of the dead female (and possibly of the father) from a previous year's litter. More interestingly, the biologists placed the remaining two orphaned pups in the den of an unrelated pair that had only two pups of their own. The orphaned pups were accepted and raised by the pair along with their own pups. In another example, an orphaned pup was introduced to a mother in captivity with four pups. The orphaned pup was accepted by the family group, and upon release from captivity, the pup remained a part of the family.

A mated pair of adult foxes as well as any young that have not yet dispersed continue to associate even after the breeding season and pup-rearing are over. Members of a family group commonly share dens and occasionally forage together. The male and female of a mated pair will share dens throughout the year. Non-dispersing offspring will continue to share dens with their parents on occasion at least to 18 months of age, and siblings have been found sharing dens up to 21 months of age (Koopman et al. 1998). Kit foxes have also occasionally been found sharing dens with individuals in adjacent home ranges, if those individuals are related (Ralls et al. 2001; Ralls and White 2003). Thus, kit foxes can maintain relationships with relatives even after they disperse, if the dispersing individuals end up in an adjacent home range.

Dispersal tends to be male-biased (see Space Use and Movements). A higher proportion of males disperse and also tend to go farther from their natal home range. Females seem to be more likely to delay dispersal, and when they do disperse, they do not travel as far from the natal home range. Not uncommonly, they may move one or two home ranges away. Females in a given area tend to be more closely related than females farther away. This results in somewhat of a matriarchal structure to local populations. It also may explain why males disperse farther away, as this may be a strategy to avoid inbreeding.

BEHAVIOR

San Joaquin kit foxes, and kit foxes and swift foxes in general, exhibit no behaviors that would be considered extraordinarily different from those of other canids. Vocalization is a rare behavior among kit foxes, and it may be an adaptation to avoid attracting larger predators, particularly coyotes. Even young pups rarely vocalize. Observing a litter of pups vigorously playing at a natal den is almost a bit surreal because of the virtual absence of sound, despite their frenetic ambushing, chasing, biting, tail-pulling, and wrestling. Yet, no sounds can be heard. It is like watching a silent movie!

Kit foxes have been heard to make four types of sounds on rare occasion. During the breeding season, particularly in December when the females are in estrus (Murdoch et al. 2008a), adult kit foxes sometimes emit a call that appears intended to locate mates. Mostly it seems to be used by males, but females have been observed emitting it as well. Sometimes it brings members of an established pair together, and sometimes it brings together unpaired individuals, which could result in the establishment of a new pair or an extra-pair copulation (see Social Ecology). Murdoch et al. (2008a) described this vocalization as a bark, although it can also sound somewhat like a short scratchy scream. Egoscue (1962) referred to it as the "lonesome call," and in closely related swift foxes it has been referred to as a "lubricious bay" (Conover 2001).

Kit foxes can also make a soft "chittering" sound. This seems to be a sign of submission as it made mostly by pups or younger foxes in the presence of adults, or sometimes by a female toward her mate. Interestingly, a captive kit fox at the Fresno Chaffee Zoo was observed making this sound when a favorite keeper who sometimes brought treats approached the enclosure. This chittering sound has also been heard in an aggressive or defensive context when foxes from neighboring home ranges encounter each other near a home range boundary.

Kit foxes will give a sharp alarm bark when any potential threat approaches a natal den, and this can morph into a high-pitched aggressive scream if the intruder approaches too closely. Clearly, this vocalization is intended to intimidate and hopefully drive away the intruder. We have observed this when inspecting natal dens as part of our research and also when a dog, cat, or red fox have come near natal dens. Not surprisingly, some kit foxes will also emit a similar scream along with much growling when they are in a trap and a biologist approaches.

Finally, kit foxes will emit a soft closed-mouth call that sounds like "roop?"; this appears to be an anxiety vocalization. Foxes in traps or handling bags make this sound and it is heard occasionally from foxes down in dens when intruders are outside the den. Others have described this sound as a croaking noise (Egoscue 1962) or the sound of a perking coffee pot (Morrell 1972).

Kit foxes have not been observed displaying much intraspecific aggression. They do not appear to be fiercely territorial, unlike some other canids (e.g., island foxes, wolves) where encounters between neighboring animals can result in fighting, injuries, and even deaths. Resident kit foxes have been observed chasing intruding kit foxes out of a home range area (Murdoch et al. 2008b), but most intraspecific spacing appears to be achieved through scent-marking. Kit foxes scent-mark extensively as they travel around their home ranges (Murdoch 2004). Typical of other canids, the marks consist largely of a few drops of urine commonly placed on objects to facilitate detection by conspecifics, including rocks and vegetation along travel routes (Murdoch 2004). Also, typical of many carnivores, kit foxes will deposit scats in latrines (Figure 2.32), some of which can be quite large. Kit foxes and coyotes commonly mark each other's fecal deposits, resulting in latrines consisting of scats of both species. More unique to kit foxes, they will also sometimes mark objects with small fecal depositions. Kit foxes particularly like to mark novel objects in their environment, possibly because they know that any other fox passing by will also want to investigate that object. As one example, kit foxes frequently mark small mammal traps (Figure 2.33).

Hunting behavior of kit foxes seems to consist primarily of searching in areas where prey are potentially present and then opportunistically encountering and capturing prey that are aboveground. Kit foxes have not been observed hiding and waiting to ambush prey. Nor have they been observed trying to excavate prey. Instead, there is typically a very short rush or dash, and prey is either captured or missed and the fox moves on and continues searching. As mentioned, kit foxes are kangaroo rat specialists and seem especially adept at capturing this prey.

Kit foxes have been observed to cache food on occasion. Most prey are small enough that they are typically consumed whole (e.g., small rodents, lizards, invertebrates), and so usually there is not much left to cache. However, foxes have been observed caching small bits of prey around natal dens. Most caching has been observed among urban kit foxes, where food surpluses are more likely to be encountered. Small bits of food tossed to a fox were observed to be cached, and then the fox would return to obtain more. In some interesting caching observations, on a number of occasions foxes were seen caching golf balls they were picking up from golf course driving ranges. They usually picked up a ball, went a few meters, dug a shallow excavation, dropped the ball in, pushed dirt over it with their nose, and then repeated this with other golf balls. A best guess is that the balls may have resembled eggs, and despite not tasting or

FIGURE 2.32. A San Joaquin kit fox fecal latrine. Photo by Brian Cypher.

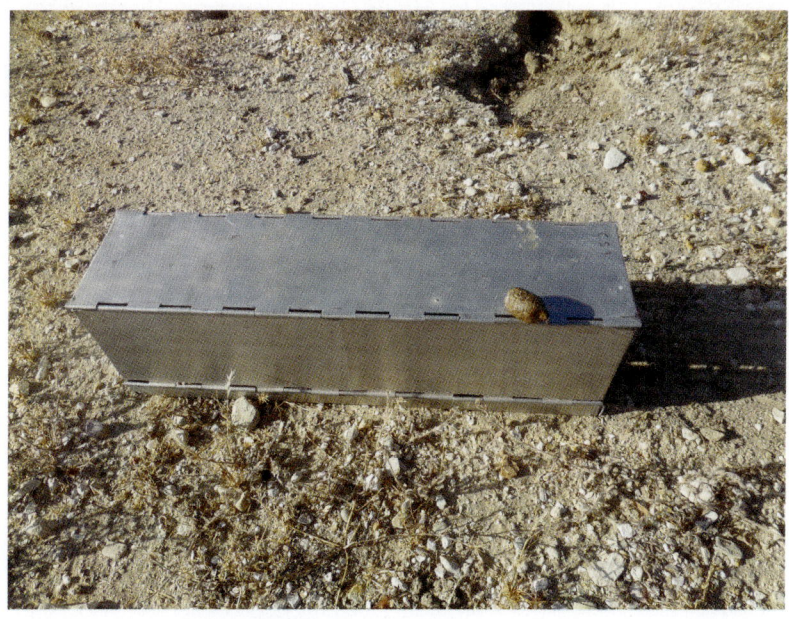

FIGURE 2.33. San Joaquin kit fox fecal deposit on a Sherman small mammal trap. Photo by Brian Cypher.

smelling like an egg, such caching may have been a deeply ingrained instinctive behavior. Arctic foxes commonly cache bird eggs (Careau et al. 2008), and red foxes have been observed caching eggs as well (Macdonald et al. 1994).

One final hunting behavior that is quite interesting is that kit foxes have been observed following badgers. Presumably, the foxes were waiting for the badgers to begin attempting to excavate a burrowing rodent, and then hoping to catch the prey as it exited out of another burrow entrance (Clark et al. 2015). This behavior has been well documented among coyotes (e.g., Minta et al. 1992) and has also been observed among swift foxes (Ausband and Ausband 2006). The frequency of successfully obtaining a meal versus foraging independently is unknown.

Another interesting aspect of kit fox behavior is one that is not necessarily obvious. As with many species (Wilson et al. 1994), individual kit foxes fall along a spectrum of "shy/bold" behavior. Thus, some individuals are more curious, less wary, and more likely to investigate novel objects. Others are more wary and cautious and inclined to avoid strange objects. In work conducted by Bremner-Harrison and colleagues (Bremner-Harrison and Cypher 2011; Bremner-Harrison et al. 2013), they found that foxes in natural environments tended to be more wary with regard to any potential threats, in this case a toy coyote that also emitted recorded howls, and less inclined to investigate novel objects, including foods not found naturally in their environment (in this case, chopped crab meat). Foxes in urban environments also exhibited wariness of potential threats but were much more likely to investigate novel objects and novel foods. This makes perfect sense, as urban foxes likely frequently encounter new potential foods, particularly those of an anthropogenic origin (e.g., pet food, discarded food, trash). There may be some selection in urban environments for bolder foxes that will investigate novel foods as well as novel denning opportunities (e.g., landscape planters, pipes, rubble piles, underneath storage containers and buildings). Exploiting these resources and opportunities likely enhances their ability to thrive in this novel environment.

INTERSPECIFIC INTERACTIONS

Kit foxes interact with a number of other species in a variety of ways. Predation is an obvious interaction and has been addressed elsewhere. To briefly summarize, kit foxes constitute the predator in interactions with a wide variety of species (see Foraging Ecology). Kit foxes also constitute prey for some species such as bobcats and golden eagles (see Survival and Mortality Factors). However, the discussion here will focus on interactions other than predation. Most of these constitute competitive interactions categorized as either interference competition or exploitative competition. In addition to predation, interference competition includes killing, harassing, and excluding another species in an effort to reduce competition for resources (Case and Gilpin 1974; Pianka 1978). Exploitative competition does not necessarily involve any direct physical interactions between two species, but instead involves the consumption or acquisition of resources by one species such that they are not available for use by another species (Pianka 1978). Kit foxes engage in both types of competition with other species, and they also interact with some other species in a manner that does not fall into either of these categories.

Coyotes are a primary source of mortality for San Joaquin kit foxes throughout most of their range (see Survival and Mortality Factors). Coyotes rarely consume the kit foxes they kill, although consumption of carcasses apparently increases during times like extended drought when food resources are more limited (Ralls and White 1995). Instead, mortality from coyotes is better described as interference competition and is commonly more specifically characterized as intraguild competition. Guilds are loosely defined as collections of species with similar ecologies such that their use of resources overlaps, thereby making them competitors with each other (Polis et al. 1989; Johnson et al. 1996; Linnell and Strand 2000). In such interactions, one competitor is commonly dominant over the other, usually by virtue of larger size. This is exactly the situation between coyotes and kit foxes, and indeed between coyotes and all other North American foxes (Cypher 2003). Coyotes and foxes use many of the same food resources, and so exploitative competition is occurring (Cypher and Spencer 1998). Killing foxes is a strategy by coyotes to reduce this competition. Based on evidence to date, coyotes do not seem to actively hunt kit foxes, but if they encounter them and have an opportunity to kill a kit fox, they will apparently do so. One exception to the hunting generalization is that coyotes are commonly seen visiting kit fox natal dens when young pups are present. Coyotes could be trying to prey on the pups, or they may also just be attracted to the dens by the abundant activity and scent from the foxes as well as by odors from decomposing prey remains.

The relationship between kit foxes and other fox species is also best described as intraguild competition. Kit foxes generally do not overlap spatially with other foxes, although there are a few exceptions. Kit foxes may overlap with gray foxes to a limited extent, mostly around the margins of suitable kit fox habitat. These include areas where the more arid scrublands and grasslands preferred by kit foxes begin transitioning to chaparral or oak woodland savanna habitat. Both kit foxes and gray foxes have been detected within these transition zones. Gray foxes are larger than kit foxes (5.0 kg vs 2.3 kg or 11.0 lb vs 5.1 lb), although to date, no deaths of kit foxes attributable to gray foxes have been documented. However, the two species likely use some of the same food resources in these overlap areas, and so some exploitative competition may be occurring.

Red foxes are also larger than kit foxes (5.5 kg vs 2.3 kg or 12.1 lb vs 5.1 lb). However, historically the two species did not overlap at all. In North America, red foxes were found primarily in boreal and montane habitats (Kamler and Ballard 2002). In California, red foxes were found only in the Sierra Nevada and southern Cascade ranges (Grinnell et al. 1937; Lewis et al. 1999), although there is some contention that they may also have occurred naturally in parts of the Sacramento Valley (Sacks et al. 2010). However, they did not occur anywhere in the San Joaquin Desert, Mojave Desert, or arid southern California valleys, and therefore did not overlap with any of the subspecies of kit foxes in California. In the late 1800s and early 1900s, red foxes that probably were European in origin were brought to California for fur farming and possibly hunting (Grinnell et al. 1937; Lewis et al. 1999). Many animals escaped from both ventures and generally found California quite suitable. Thus, non-native red foxes have not only become established in California, but they have also spread extensively throughout the state and are thriving. Red foxes generally prefer more mesic habitats (Cypher 2003). Presently, most mesic habitats in the San Joaquin Desert (e.g., lake margins, wetlands, sloughs, riparian areas) have been drained and converted to other uses, usually agricultural. However, this irrigated agriculture along with the many rapidly expanding urban environments in the San Joaquin Valley provide an abundance of water sources, and red foxes are present throughout these areas.

Red foxes are rarely seen in intact arid habitats within the range of the San Joaquin kit fox. This absence is due partly to a lack of water but is due also to intense interference competition from coyotes (Cypher, Clark, et al. 2001). Red foxes do not use daily dens as do kit foxes or climb trees as do gray foxes. Thus, they are more easily killed by coyotes, and this pressure significantly reduces and even excludes use of natural arid habitats by red foxes. Therefore, red foxes are found primarily in agricultural and urban areas where both water and cover are abundant and where coyotes are rare or absent because of anthropogenic disturbance and harassment. Consequently, kit foxes encounter red foxes only in some urban areas (see Kit Foxes and Urban Areas) or near the interface of agricultural and natural lands. In an interesting situation, both kit foxes and red foxes were present in the narrow band of habitat along the California Aqueduct in the Lost Hills area in Kern County. Kit foxes appeared to mostly avoid the red foxes, but one kit fox was suspected to have been killed by a red fox (Clark et al. 2005). Also, a kit fox on the Carrizo Plain was reportedly killed by a red fox (Ralls and White 1995). However, kit fox mortality from red foxes is clearly rare. Red foxes do engage in exploitative competition with kit foxes by consuming many of the same foods used by kit foxes (Cypher, Deatherage, et al. 2022) and also by using kit fox dens on occasion (Figure 2.34).

Interactions between kit foxes and domestic dogs and cats also potentially fall into the realm of intraguild competition. Dogs in natural lands, likely associated with nearby residences, have killed San Joaquin kit foxes (Spiegel 1996), as have dogs in urban environments (Cypher 2010). In both situations, the fox carcasses are typically not consumed. This killing is likely an instinctive remnant from pre-

FIGURE 2.34. A red fox entering a San Joaquin kit fox den on the campus of the California State University-Bakersfield. Photo by Alyse Gabaldon.

domestication in the evolution of dogs, just like the propensity of dogs to chase after any small prey–resembling animal they encounter. Interactions between kit foxes and domestic cats (feral or pets) are different. Structurally, both species are similar in size, although cats typically outweigh the foxes. Without a clear advantage to either species, the typical interaction is avoidance. Kit foxes and feral cats may even use dens in close proximity at times. in what appears to be a case of coexistence. However, at feeding stations for feral cats in Bakersfield, kit foxes commonly visit for food as well but defer to cats (Harrison et al. 2011).

Interesting interactions occur between kit foxes and badgers. The behavior of kit foxes following badgers and attempting to capture prey escaping from badgers was described (see Behavior). Technically, this is a type of kleptoparasitism on the part of the kit foxes; however, kit foxes and badgers occasionally compete for dens. Kit foxes will sometimes expand badger diggings into a den. In addition, badgers apparently sometimes take over kit fox dens. In one instance described from Camp Roberts in San Luis Obispo County, a kit fox and a badger apparently fought for possession of a den resulting in the death of the kit fox (Standley et al. 1992).

Interesting interactions also occur between kit foxes and burrowing owls. These owls commonly use burrows and dens of other species in the San Joaquin Desert. They will use California ground squirrel burrows but also occasionally use vacant kit fox dens. On a number of occasions, kit foxes have been observed using a den for some days, and when they move to a different den, burrowing owls move into the first den. The owls use it for some time, and then kit foxes may move back in. In an exceedingly interesting situation, on four occasions in the urban environment of the city of Bakersfield, burrowing owls were observed entering dens in which a radio-collared kit fox was present (ESRP, unpublished data). It is unknown whether kit foxes ever kill burrowing owls, although the owls have been observed dive-bombing and chasing kit foxes on several occasions.

Striped skunks are present throughout the range of the San Joaquin kit fox and are quite abundant in urban areas, including those inhabited by kit foxes. Kit foxes generally avoid the skunks, as do most animals. Some exploitative competition may occur between the two species, but it is likely not extensive. On occasion, skunks have been observed using known kit fox dens, particularly in urban areas (Figure 2.35). During a study of skunk-fox interactions in Bakersfield, not only did foxes and skunks use the same dens, but on four occasions, radio telemetry revealed that collared skunks and collared kit foxes were also using the same den simultaneously (Harrison et al. 2011). This presumably platonic coexistence has a significant down-side, that being the potential for disease transmission. As previously mentioned, a significant decline in kit fox abundance in the Salinas Valley was coincident with a rabies epidemic in striped skunks (White et al. 2000), and den sharing could have potentially been a significant contributor to the interspecific transfer of the virus.

Finally, in what is likely a commensal relationship with no effect on kit foxes, some number of species likely use kit fox dens for shelter, probably including both occupied and unoccupied dens. Many of

FIGURE 2.35. A striped skunk entering a San Joaquin kit fox den in Bakersfield. Photo by Alyse Gabaldon.

these species are probably invertebrates but also include some vertebrates. Beetles and scorpions have been observed in the entrances to kit fox dens. In addition to the red foxes and burrowing owls as mentioned above, California ground squirrels and side-blotched lizards have been observed entering kit fox dens, and Pacific rattlesnakes have been observed lying just inside the entrance.

Chapter 3

CONTEMPORARY CHALLENGES

A number of factors led to San Joaquin kit foxes becoming rare and eventually being granted federal and state protections (US-FWS 1998). Some of these factors, such as harvests and predator control programs, no longer constitute significant threats. Habitat loss was another important factor in the decline of San Joaquin kit foxes and continues to pose a threat. Various land uses have contributed to this loss. Interestingly, San Joaquin kit foxes respond to these land uses in different ways, and not all uses exclude continued use of the land by kit foxes, particularly if certain mitigation measures are implemented. In this chapter, various common land uses within the range of the San Joaquin kit fox are discussed, including the response by kit foxes to these uses. The chapter concludes with a discussion of climate change, the effects of which may potentially be very different for San Joaquin kit foxes compared with those predicted for many other species.

KIT FOXES AND OIL FIELDS

The extraction of crude oil and natural gas is a common land use in a significant proportion of the range of the San Joaquin kit fox. This activity consists primarily of drilling wells to reach underground reserves of hydrocarbon resources, and then bringing these resources to the surface and transporting them to storage or processing facilities. In the case of crude oil, some sort of pumping unit is typically installed on the well to facilitate the flow of oil. Well pads are commonly 0.2–4 ha (0.5–10 ac) in size. Additional infrastructure includes access roads, electrical transmission lines (to power pumping units), pipelines to transport oil and gas, storage tanks, and processing plants, among others. The amount of land disturbance depends on the density of wells and facilities in a given area. In low-density oil fields, less than 10% of the habitat may be disturbed, while in high-density oil fields, very little habitat may remain. Drilling for new wells may be conducted around the clock. Otherwise, most oil field activity occurs primarily during daylight hours and with little human activity after dark, although pumping units and processing facilities can make considerable noise on a continuous basis.

San Joaquin kit foxes are usually present in most oil fields within their range (e.g., Spiegel 1996; Cypher et al. 2000; Fiehler et al. 2017). Although oil field operations reduce the total amount of available habitat, in most fields considerable habitat is usually present between the roads and facilities. These patches commonly harbor prey species such as kangaroo rats, pocket mice, deer mice, ground squirrels, rabbits, lizards, birds, and various invertebrates. The patches also provide cover for kit foxes and areas where foxes can dig dens. Kit foxes also readily establish dens under buildings and storage containers (e.g., seatrains), and within culverts and stored pipes. Hazards for kit foxes are certainly present in oil fields and include impacts with heavy equipment and other vehicles, toxic liquids and gases, and oil spills (Figure 3.1). Despite these hazards and risks, kit foxes continue to occupy and use oil fields.

Since passage of the Endangered Species Act in 1973, various mitigation measures have been routinely implemented in oil fields to avoid or minimize impacts to kit foxes (see Feinstein et al. 2015 for a detailed description). Of particular importance has been the permanent conservation and management of habitat in compensation for the impacts of oil field activities. Some examples include more than 2400 ha (6000 ac) of habitat conserved in the Coles Levee Ecopreserve by ARCO, more than 3200 ha (8000 ac) conserved in association with the Elk Hills oil field, and 340 ha (840 ac) conserved by Nuevo-Torch in the Lokern area. All these lands are conserved and managed in perpetuity for San Joaquin kit foxes and co-occurring rare species.

FIGURE 3.1. San Joaquin kit fox that has been covered in oil from a spill.
Photo by Tory Westall.

The California Energy Commission studied kit foxes in a natural
area and a nearby high-density oil field where habitat disturbance
was about 70% (Spiegel 1996). Despite the substantial habitat loss
and associated disturbance of oil production activities, kit fox density in the oil field was still about 50% of that in the natural area. In
a long-term study of oil field effects on kit foxes at the Elk Hills and
Buena Vista Naval Petroleum Reserves, various demographic and
ecological attributes of kit foxes were compared between areas with
and without oil field activities (Koopman et al. 1998; Cypher et al.
2000; Zoellick et al. 2002). No significant differences were found.
Habitat disturbance was about 26% in areas with oil production activities and about 8% in nonproduction areas. Kit fox numbers fluc-
tuated markedly during the 15-year study, but these fluctuations
were similar in the areas with and without oil field activities and
were a function of prey availability that varied with annual precipitation. In a comparison of various demographic and ecological parameters between kit foxes in oil fields and those in natural lands,

TABLE 7. *Comparison of demographic and ecological parameters between San Joaquin kit foxes using oilfields and those using natural lands*

Parameter	Oilfield site	Natural site	Source
Annual survival[a]	58%	52%	Spiegel 1996
	57%	38%	Cypher et al. 2000
Reproductive success[b]	50%	64%	Spiegel 1996
	69%	51%	Cypher et al. 2000
Home range size[c]	4.8 km^2	6.7 km^2	Spiegel 1996
	3.1 km^2	2.8 km^2	Zoellick et al. 2002
Dens per year[d]	13.6	15.6	Spiegel unpubl. data
	11.4	12.6	Koopman et al. 1998
Mean mass (kg)			
Adult male	2.37	2.53	Spiegel 1996
Adult female	2.26	2.11	Spiegel 1996
Juvenile male	1.83	2.03	Spiegel 1996
Juvenile female	1.81	1.76	Spiegel 1996

[a]Probability of surviving for 1 year.
[b]Proportion of adult females verified with litters.
[c]95% minimum convex polygon home range.
[d]Mean dens used per year by each adult fox.

in most cases the parameters did not differ statistically between oil fields and natural lands (Table 7). Where there was a significant difference (e.g., adult female mass), the foxes in the oil field were actually faring better.

So, thanks to their ecological plasticity and adaptability, kit foxes are not excluded from areas where hydrocarbon extraction is occurring. Consequently, more kit foxes and kit fox populations are present than if kit foxes had been excluded by this land use. Also, of potentially significant importance in the future, the hydrocarbon reserves will eventually be depleted (or rendered obsolete by emerging technologies and economics). If and when this happens, many lands currently used for hydrocarbon extraction will likely revert to habitat. It is possible that the lands could be used for alternate activities, but soil and other contamination issues will render these lands unsuitable or undesirable for many uses, such as residential or agricultural development. This could result in large areas with reasonably high habitat quality becoming available for San Joaquin kit foxes, as well as associated species. Reversion to natural habitat has already occurred on some oil field lands near Taft and Coalinga.

KIT FOXES AND SOLAR FARMS

Photovoltaic solar power energy generation has expanded rapidly worldwide (REN21 2022) and particularly in California (SEIA 2016). Solar energy development was facilitated by various federally legislated economic incentives. In California, further incentive was provided by bills passed by the California legislature that mandated increasing levels of energy production from renewable energy sources with the latest bill requiring that all power-supplying utilities obtain 100% of their electricity from such sources by 2045 (Cameron et al. 2012; Pearce et al. 2016; Cypher, Boroski, et al. 2021). As of 2019, 748 solar plants were operating in California with many more planned for construction (CEC 2020; KCPD 2014).

The San Joaquin Desert region has been a focal area for photovoltaic solar energy development due to an abundance of mostly flat terrain, high insolation rates, vegetation that is sparse and low in structure, and relatively low land prices (Butterfield et al. 2013; Pearce et al. 2016; Hoffacker et al. 2017; Phillips and Cypher 2019). Much of the highly suitable San Joaquin kit fox habitat within this region is also highly suitable for solar energy plants (Phillips and Cypher 2019). The considerable acreages typically required to construct a solar plant certainly constitute a potential threat to San Joaquin kit foxes and other species. Just during 2010–2020, six large solar facilities totaling 5091 ha (12,580 ac) were constructed in kit fox habitat (Cypher, Boroski, et al. 2021).

However, as a result of permitting requirements from the California Department of Fish and Wildlife and the US Fish and Wildlife Service and with input from species experts, a number of conservation measures have been incorporated into the design and operation of the facilities constructed in kit fox habitat. These measures are intended to facilitate kit foxes and other species in continued use of the lands where solar facilities are built. Two measures in particular have been critically important in facilitating use of the facilities by kit foxes: one is permeable security fencing, and the other is the retention and management of vegetation within the facilities. At each of the facilities, the security fence surrounding the arrays of solar panels is designed to be permeable to kit foxes. The fences are typically 2.4 m (7.9 ft) tall, sometimes with strands of barbed wire on the top. At most facilities, the fencing used was 5 cm (2 in) mesh chain-link. To make it permeable to kit foxes, a gap of approximately 12–15 cm (5–6 in) was left between the bottom of the fence and the ground (Figure 3.2). Kit foxes can easily pass through this gap. At other facilities, a deer-proof style fence with 15×15 cm (6×6 in) mesh openings was used (Figure 3.2). All these designs allow kit foxes and any similar-sized or smaller species (such as kit fox prey species) to pass through the fences and freely enter or exit the facilities. They also provide an added benefit in that they inhibit passage by larger species such as coyotes and bobcats, both of which are potential predators of kit foxes.

The other important conservation measure was encouraging vegetation through seeding (Althouse and Meade, Inc. 2010; HTH 2012) or allowing to recover naturally in the solar panel arrays after construction was completed. Furthermore, vegetation structure on the facilities is typically managed through sheep and cattle grazing, sometimes supplemented with mechanical mowing within the arrays. The goal is to keep the vegetation structure low (ideally ≤15 cm or ≤6 in), a condition favored by kit foxes (Cypher et al. 2013), their prey, and the other co-occurring species of conservation concern.

In addition to permeable fencing and vegetation management, a number of other conservation measures beneficial to kit foxes and the other species have been implemented (Cypher, Boroski, et al. 2021). Animal movement corridors were incorporated into the design of all facilities greater than 500 ha (>1200 ac) in size. Instead of constructing the solar panel arrays in a single or a few large blocks, the arrays were distributed among a larger number of smaller groupings such that habitat corridors were maintained through the project sites. Surveys are conducted for kit fox dens prior to construction and prior to any ground-disturbing maintenance activities. Dens that can be avoided are left intact, even if temporarily covered, to facilitate continued use after construction or maintenance activities. On most of the solar sites, artificial dens have been installed for kit foxes. Other measures include prohibitions on pet or feral dogs and

FIGURE 3.2. Fences around solar facilities that have been modified to facilitate passage by kit foxes. Photos by Larry Saslaw and Brian Cypher.

firearms, rodenticide restrictions, and trash abatement programs. Speed limits (usually 15–25 kph or 9–15 mph) are strictly enforced, and off-road driving is restricted. Hazardous substance spills are rapidly cleaned up. Another common measure is that employees, contractors, and others working outdoors at the facilities are required to complete an environmental awareness program, including recognition of kit foxes and other species of conservation concern. Finally, agency-approved "designated biologists" with kit fox expertise provide input on activities that could potentially cause harm to species of conservation concern. As a result of these conservation efforts, San Joaquin kit foxes, along with most of the other species of conservation concern present on or near these San Joaquin Desert solar sites prior to construction of the facilities, are still present.

Two particularly large solar energy plants were constructed on the north end of the Carrizo Plain region. Much of this region is considered a core area for the conservation and recovery of San Joaquin kit foxes (see Distribution and Habitat Preferences). Thus, the location of these proposed plants was a concern. The Topaz Solar Farms (TSF) plant had a 1902 ha (4700 ac) footprint, while the California Valley Solar Ranch (CVSR) plant had a 797 ha (1970 ac) footprint. In addition to the potential loss of habitat, there was a concern that the solar farms would impede kit foxes from entering a corridor on the north end of the Carrizo Plain that connected the region to habitat to the east in Kern County. Additionally, another large solar plant, the Panoche Valley Solar Farm (PVSR), was constructed in the Panoche core area. The PVSR had a 504 ha (1022 ac) footprint. The authorization permits issued for construction of the plants required all the typical impact avoidance and minimization measures, including the modified security fencing described earlier. Thus, kit foxes would be able to access both sites and use the lands therein or pass through. To further mitigate for impacts to kit foxes, approximately 22,788 ha (56,400 ac) of nearby high-quality habitat was conserved and transferred to conservation organizations (i.e., Center for Natural Lands Management, Sequoia Riverlands Trust, California Department of Fish and Wildlife) along with endowments for management. The permits also required that the owners of each plant fund a three-year post-construction study at each facility to assess the effects of the plants on kit foxes. Construction of the Carrizo plants was completed in the fall of 2014, and construction of the Panoche Valley plant was completed in winter 2019. Kit fox studies were completed at the Carrizo sites in winter 2017 and at the Panoche Valley site in spring 2022.

The study designs for all the research projects were similar (Cypher, Boroski, et al. 2021; Cypher et al. 2023). Foxes on or within 1.5 km (0.9 mi, the approximate diameter of an average kit fox home range in good quality habitat; Cypher et al. 2013) of the solar sites were considered "solar foxes," and foxes residing outside these areas were considered "reference foxes." Various kit fox demographic and ecological attributes were compared between the solar and reference foxes. In all the studies, there were few differences between these two groups of foxes (Cypher, Boroski, et al. 2021; Cypher et al. 2023). In fact, in the Carrizo studies, kit fox survival actually trended higher on the solar sites. This was attributed to the fencing that allowed passage by kit foxes but significantly impeded entry by coyotes. Additionally, the solar photovoltaic panels afforded the foxes cover from attack by golden eagles, which killed a number of reference kit foxes during the study (Cypher, Westall, et al. 2019; Cypher, Boroski, et al. 2021a). Kit foxes on the TSF and PVSR solar study areas exhibited larger home ranges and greater movements and den switching compared with foxes on the reference areas (Cypher, Westall, et al. 2019; Cypher et al. 2023). However, in the case of TSF study, these were attributed to the fact that the solar plant was built primarily on lands that had been dryland farmed up until construction of the plant was initiated. Because of this previous disturbance, both prey and dens were not abundant, which resulted in greater space use and movements. At the PVSF solar site, fewer prey were likely available, due both to disturbance associated with construction and also to the fact that almost 700 giant kangaroo rats, the main prey for kit foxes, were translocated off the site prior to construction. Kit foxes apparently traveled off the site to forage but returned to site for denning.

In all three of the studies, kit foxes used the solar site extensively, both during construction as well as after construction was completed. Indeed, the continued presence of kit foxes on the sites during construction created issues that occasionally impeded construction activities. These included kit foxes seeking cover under equipment

FIGURE 3.3. San Joaquin kit fox (a) in construction materials and (b) under a construction vehicle on a solar farm. Photos by Jason Dart.

FIGURE 3.4. San Joaquin kit fox relaxing in the shade of solar panels on the Topaz Solar Farms. Photo by Christine Van Horn Job.

or construction materials (Figure 3.3), building dens in active construction areas or inconvenient locations (e.g., next to facilities or roads), and raising litters of pups in or near construction areas. Demographic and ecological data from the studies indicated the kit foxes on the solar sites were functioning in a manner similar to foxes on nearby reference sites (Table 8). Furthermore, demographic and ecological attributes were similar between solar study sites and non-solar study sites with high-quality habitat in core population areas (Table 9). Clearly, solar facilities can be constructed and operated in a manner compatible with continued use by kit foxes (Figure 3.4). This point was highlighted in an analysis by Phillips and Cypher (2019), who also described the potential habitat enhancement effect of solar sites constructed in marginal kit fox habitat and also suggested that solar plants could be strategically located to connect fragmented blocks of habitat.

Continued use of appropriately mitigated solar sites by San Joaquin kit foxes has also been observed on at least three other solar facilities: the 1174 ha (2901 ac) California Flats Solar Plant, the 567 ha

TABLE 8. *Comparison of demographic and ecological values for San Joaquin kit foxes on solar sites and associated reference sites in the San Joaquin Desert region, California*

Kit fox attribute	Solar site	Reference site
Probability of survival		
TSF	0.65	0.49
CVSR	0.76	0.66
PVSF	0.66	0.75
Reproductive success (%)		
TSF	100	88.9
CVSR	86.7	86.7
PVSF	92.9	81.3
Mean litter size (range)		
TSF	4.3 (2–8)	3.9 (1–7)
CVSR	3.2 (1–5)	4.5 (2–7)
PVSF	3.4 (1–5)	4.0 (1–7)
Mean mass (kg): males		
TSF	2.48	2.64
CVSR	2.69	2.53
PVSF	2.67	2.67
Mean mass (kg): females		
TSF	2.16	2.16
CVSR	2.22	2.15
PVSF	2.25	2.17
95% MCP home range (km^2)		
TSF	9.4	5.1
CVSR	3.9	4.2
PVSF	6.1	2.4
Mean dens per fox		
TSF	11.2	8.4
CVSR	15.1	19.4
PVSF	–	–

Data sources: TSF—Cypher, Westall, et al. 2019; CVSR—HTH 2019; PVSF—Cypher, Westall, et al. 2023.
Note: Solar sites and years that kit foxes were studied: TSF = Topaz Solar Farms (2014–17); CVSR = California Valley Solar Ranch (2014–17); PVSF = Panoche Valley Solar Farm (2019–22). See text for detailed definitions of the attributes.

TABLE 9. *Comparison of demographic and ecological values for San Joaquin kit foxes on solar and non-solar sites in population core areas in the San Joaquin Desert region, California*

Kit fox attribute	Solar sites	Non–solar sites
Probability of survival	0.65–0.76	0.38–0.83
Reproductive success (%)	86.7–100	0–100
Mean litter size (range)	3.2–4.7	2.0–5.4
Litter size range	1–8	1–9
Mean mass (kg): males	2.48–2.69	2.52–2.67
Mean mass (kg): females	2.16–2.25	2.11–2.2
95% MCP home range (km^2)	3.9–9.4	1.3–11.4
Mean dens per fox	11.2–15.1	8.4–19.4

Data sources for solar sites: Cypher, Westall, et al. 2019, 2023; HTH 2019.
Data sources for non-solar sites: Cypher et al. 2009, 2014, 2000, 2019*b*, 2022; H.T. Harvey and Associates 2019; Koopman et al. 1998; Nelson et al. 2007; Ralls and White 1995; Spiegel 1996, unpublished data; Warrick and Cypher 1999; White and Ralls 1993; Zoellick et al. 2002.

(1401 ac) Wright Solar Plant, and the 125 ha (309 ac) Hills/Blackwell Solar Farm. Additionally, in all cases, considerable high-quality habitat has been conserved as mitigation for these facilities, and these lands along with sizeable endowments have been transferred to conservation organizations for management in perpetuity for kit foxes and other species. Thus, solar plants constructed in the range of the San Joaquin kit fox have not had significant negative impacts and in most cases have even had some positive effects. This result is quite encouraging, as many more solar energy generating facilities are likely to be constructed in the San Joaquin Desert region, particularly in the next two decades as landowners seek alternative uses for lands retired from agriculture because of increasingly limited groundwater availability (see Kit Foxes and Agricultural Lands).

KIT FOXES AND AGRICULTURAL LANDS

The San Joaquin Valley is actually a rather fascinating region from the perspective of agricultural production (Figure 3.5). Because of relatively fertile soils and mild climate, more than 250 different crops are grown in the valley, with the major ones being tree nuts and fruits, table and wine grapes, citrus, hay, cotton, tomatoes, and

FIGURE 3.5. Aerial view of agricultural lands in the San Joaquin Valley, California. Photo by Brian Cypher.

cereal grains. The net worth of these crops as of 2022 was approximately $17 billion per year. Approximately 25% of the nation's food is produced in the valley, including 40% of the fruits, nuts, and other table foods (USGS 2021). The agricultural productivity and importance of this region is undeniable, but it comes with a high environmental cost.

The story on the effects of agricultural activities on San Joaquin kit foxes is much different from that of oilfields and solar farms. Indeed, conversion of natural lands for agriculture is the primary factor responsible for the endangerment of the San Joaquin kit fox, along with a number of co-occurring species in the San Joaquin Desert (USFWS 1998). Kit foxes have an extremely limited capacity to use most agricultural lands. These lands are frequently disked (plowed), and this ground disturbance destroys the dens of kit foxes. It also eliminates most burrowing rodents, which comprise a significant portion of the diet of San Joaquin kit foxes. Similarly, the flood type irrigation practice commonly used for many crops also precludes den establishment. Thus, kit foxes have little if any opportunity to create the dens so crucial for their survival, and prey abundance is low. Prey are made even scarcer by the routine use of pesticides in agricultural areas to control rodents and insects. Carcasses of dead rodents in particular constitute a hazard for kit foxes due to toxic residues. Secondary poisoning of kit foxes from consumption of rodenticide-killed rodents has been documented (Huffman and Murphy 1992; Standley et al. 1992; Cypher, McMillin, et al. 2014). Other chemicals routinely used in agricultural areas, such as herbicides, fertilizers, defoliants, and others can also be toxic to kit foxes and their prey. Thus, agricultural lands provide little habitat value for kit foxes.

In a unique study conducted in Kern County, Knapp (1978) monitored 13 radio-collared kit foxes on natural lands being actively converted to agriculture. Two of the collared foxes and an uncollared fox died when entombed in their dens during tilling, and several other foxes were displaced and subsequently dispersed. The remaining foxes either avoided or exhibited only infrequent use of the tilled areas. The reduction or elimination of preferred prey was also documented.

Avoidance of agricultural lands has been evident in a number of kit fox telemetry studies conducted in close proximity to such lands. In a study carried out in northern Kern County, Warrick et al. (2007) found that kit foxes infrequently ventured up to 1 km (1.6 mi) into annual crops (e.g., cotton, barley) during nocturnal foraging. The foxes always returned to adjacent natural habitat by dawn. Kit foxes were found to occasionally spend a day within almond orchards but almost always returned to natural lands by the end of the next night. Similarly, kit foxes east of Bakersfield occasionally spent one or two days in orange groves located adjacent to natural lands (Cypher and Brown 2006). Kit foxes in the northern Carrizo Plain clearly avoided dryland farmed parcels (Cypher, Westall, et al. 2019). They occasionally crossed through these parcels but otherwise spent little time in them. In addition to the lack of dens and prey, predation risk on these parcels was elevated as little or no vegetation was present to provide cover. In telemetry studies conducted in the Semitropic area of Kern County (Cypher, Westall, et al. 2014) and in the Panoche Valley (ESRP unpublished data), kit foxes exhibited similar avoidance of actively tilled areas.

FIGURE 3.6. San Joaquin kit foxes on grazing land. Photo by Tory Westall.

Grazing lands are sometimes categorized as agricultural lands; however, the response by foxes to grazing lands, or rangelands, is completely different from their response to lands that are tilled and/ or irrigated. Grazing lands are largely intact ecologically, and all the habitat components required by kit foxes tend to be present. Human presence is infrequent, and the foxes seem undisturbed by cattle (Figure 3.6). Thus, throughout their range, kit foxes are frequently found on grazing lands. Grazing may even improve habitat quality for kit foxes by reducing vegetation height and density, thereby resulting in a more favorable vegetative structure. Jensen (1972) even suggested that overgrazing of dense grasslands might result in increased kangaroo rat abundance, and therefore, more kit foxes. A potential hazard associated with some grazing lands is the use of rodenticides, in particular for control of California ground squirrels (Cypher, Westall, et al. 2019).

Conversion of natural habitat to agricultural uses was still occurring annually within the range of the San Joaquin kit fox as of 2023. Thus, available habitat for kit foxes and co-occurring species was still declining. A considerable proportion of the agriculture in the San

Joaquin Valley is dependent on groundwater pumped from natural aquifers underlying much of the valley; however, these aquifers have been in a state of considerable and unsustainable overdraft as of 2021. In 2014, the California state legislature passed the Sustainable Groundwater Management Act (SGMA), requiring local governments and water agencies to halt overdrafting and bring groundwater basins into balanced levels of pumping and recharge. Under SGMA, these basins must reach sustainability within 20 years of the implementation of sustainability plans. For critically overdrafted basins, that will be 2040. For the remaining high and medium priority basins, 2042 is the deadline. A consequence of this act is that a considerable quantity of currently tilled land, possibly as much as half a million acres, will be retired from irrigated agriculture. Alternate uses for these retired lands include restoring some back to habitat, and this could potentially reverse the current net loss of natural habitat into a net gain (Butterfield et al. 2021); however, competing uses and a natural desire to make a profit from the lands could limit this gain.

Restoring former agricultural lands to habitat for kit foxes and co-occurring species has its challenges, and an effective approach has not been demonstrated to date. Interestingly, when it comes to restoring habitat on disturbed lands, "less" is generally better, particularly with regard to vegetation. Preferred habitat conditions (see Distribution and Habitat Preferences) include sparse ground cover and low vegetative structure with few or no shrubs. Thus, intensive seeding or planting would not be necessary. In many locations, seeding or planting may not be needed at all as seeds will simply blow in from surrounding lands. In most cases, the real challenge will be preventing vegetation from becoming too dense and tall, particularly non-native grasses (e.g., bromes, wild barley, wild oats) and other weedy species (e.g., mustards, tumbleweeds, fivehook bassia). Some form of active management (e.g., grazing) will likely be necessary to control vegetation density and structure.

A key to the restoration challenge will be ensuring that adequate prey are available for kit foxes. If restored lands are contiguous with lands occupied by prey species, there would be a reasonable likelihood that these species will eventually colonize the restored lands. However, if there is no connectivity or if there is a desire to speed up the process, then prey species could be introduced to the restored lands. Excellent candidate species would be Heermann's kangaroo rats, California ground squirrels, and pocket gophers (Cypher, Cypher, et al. 2021). All are common prey for kit foxes. All these species are habitat generalists that can live in a wide range of habitat conditions. All are also "ecosystem engineers" that modify their environment, primarily through extensive burrowing, and thus create conditions that facilitate colonization and occupation by other species, including other rodents, lizards and snakes, and invertebrates. Finally, all the candidate species are widespread and abundant, which would facilitate identifying source populations. Ground squirrels were recently successfully introduced to restored lands in southern California to enhance habitat conditions for burrowing owls (Swaisgood et al. 2019). Introducing kangaroo rats is a bit more challenging, with most attempts to date being unsuccessful (Tennant et al. 2013); however, with additional research, an effective strategy could likely be developed. Other species consumed by kit foxes that might be potential candidates for reintroductions include jackrabbits and side-blotched lizards. Invertebrates such as grasshoppers, crickets, and beetles would likely colonize restored lands on their own.

Finally, lands previously used for agriculture would be lacking in dens. One solution would be to install artificial dens (see Dens), including both surface dens that could provide quick escape cover and subterranean chambered dens for daytime resting and even reproduction (Cypher, Murdoch, et al. 2021). Kit foxes would eventually dig their own dens or modify burrows of other species (e.g., ground squirrels, kangaroo rats, badger digs); however, some initial cover would be necessary to facilitate successful occupation by kit foxes.

KIT FOXES AND ROADS

Roads constitute a potentially significant threat to kit foxes (Cypher 2000; Bjurlin and Cypher 2003; Bjurlin et al. 2005), as they do for many animal species. Mortalities can result when animals try to cross roads and are struck by vehicles. Roads can act as barriers to movements, thereby disrupting demographic and genetic flow. Roads can also enhance habitat fragmentation, and the increased

FIGURE 3.7. San Joaquin kit fox killed by a vehicle. Photo by Brian Cypher.

human access provided by roads can result in greater disturbance, dumping of trash and toxins, and many other indirect effects (Forman and Alexander 1998).

Vehicle strikes (Figures 2.10 and 3.7) occur throughout the range of the San Joaquin kit fox (Cypher 2000). However, in telemetry studies where the frequency of vehicle strikes has been quantified, this source of mortality is typically relatively low and does not seem to occur on a scale that affects population dynamics (see Survival and Mortality Factors). These relatively low mortality rates may be due at least partly to the fact that many of the roads through kit fox habitat tend to be smaller (i.e., two-lane) with relatively low traffic volumes and little traffic at night (Cypher 2000). Even in an urban population of San Joaquin kit foxes in the city of Bakersfield, where vehicle strikes were the primary cause of mortality, just 21 (5.9%) of 358 collared foxes were killed by vehicles during a study conducted from 1997 through 2004 (Bjurlin et al. 2005).

The type of road is also important. Dirt roads are common all throughout kit fox habitat. These roads are infrequently traveled by vehicles and are not a barrier for movements. Indeed, kit foxes appear to use these roads for travel, based on the presence of kit fox scats and tracks commonly found along these roads. Two-lane paved roads are also generally not a significant issue for kit foxes. Particularly because foxes are active primarily at night when traffic volume is lower, the risk of vehicle strikes is low. With the low nighttime traffic volume and the absence of median structures or other barriers to movements, kit foxes seem to easily cross two-lane roads at will. Collared foxes in the Lokern area routinely crossed the two-lane paved roads in the study area. Demographic and ecological attributes (e.g., survival, reproduction, home range size, food habits) were similar between foxes inhabiting areas adjacent to roads and those in areas more distant from roads (Cypher et al. 2009).

Larger roads (e.g., those with four or more lanes, commonly with a median and occasionally with median barriers) likely present more of a risk to kit foxes. Such roads generally have higher traffic volumes with considerable vehicle traffic after dark (Cypher 2000). Thus, the potential for vehicle strikes is higher. Median barriers are particularly problematic, as foxes (and other wildlife) attempting to cross roads can get trapped up against the barriers, further increasing the probability of vehicle strikes (Figure 3.8). In addition, these barriers impede or even preclude crossing by smaller animals like kit foxes, and this reduces genetic and demographic flow across the landscape. This disruption in gene flow can cause fragmentation of habitat and populations, thereby increasing the risk of extirpation of populations through stochastic and other processes (Frankham et al. 2017).

In recent years, there has been increasing consideration of the use of crossing structures to enable wildlife to safely cross roads. Such structures can consist of overpasses, primarily bridges, or underpasses, which can vary in size from small culverts to openings large enough to accommodate vehicles and trains. The design of these structures is critical to their potential use by kit foxes. Essentially, the larger the structure and the lower the level of multiple use (e.g., vehicle or pedestrian traffic, water flows), the greater the probability of use by kit foxes. Structures under roads can be risky for kit foxes. Structures such as pipes and culverts may be used as cover by predators of kit foxes, including coyotes and bobcats. Also, kit foxes seem averse to using any structure where they cannot see through to the

FIGURE 3.8. Remains of a San Joaquin kit fox killed by a vehicle near median barriers. Photo by Erika Noel.

other side. In a study conducted on three study sites encompassing four-lane divided highways in California, 46 structures were monitored for 12 months, but no use by kit foxes was detected (Bremner-Harrison et al. 2007). Instead, evidence suggested that foxes elected to attempt to cross over roads. Clevenger et al. (2010) documented use of culvert-style crossing structures by swift foxes in Colorado and Wyoming, but use was infrequent and, at least at the Wyoming study site, use of the crossing structures was encouraged by fencing that both impeded swift foxes from crossing over the highways and also directed them to the openings of the structures.

Kit foxes have been observed using vehicle bridges to cross over roads. Possibly, they feel more secure out in the open than in a confined space. Unfortunately, there are no dedicated wildlife overpasses anywhere in the range of the San Joaquin kit fox. Such structures would probably be used by the foxes.

Roads will likely continue as an issue for San Joaquin kit foxes. The human population continues to grow in the San Joaquin Valley, as one of the fastest growing regions in the state (PPIC 2006). As the population grows, so will infrastructure, including more and larger roads and highways.

KIT FOXES AND URBAN AREAS

Urban growth is one of the primary causes of loss and fragmentation of habitat for San Joaquin kit foxes and other species in the San Joaquin Valley (USFWS 1998). The valley is one of the fastest-growing regions in California (PPIC 2006). Fresno is the largest city in the valley, and it is the 5th most-populous city in California and the 34th most-populous city in the country. Bakersfield is the second-largest city in the valley, the 9th most-populous city in California, and the 52nd most-populous city in the country (USCB 2019). The human population in the valley is projected to double by 2040 (PPIC 2006).

Although urban growth is a serious threat to San Joaquin kit foxes due to the profound loss of natural habitat, there is a very interesting twist to this situation. As has been mentioned in other chapters, a population of San Joaquin kit foxes is present in the city of Bakersfield (Figure 3.9). It is unclear as to how long kit foxes have inhabited urbanized areas in Bakersfield. The presence of kit foxes was noted in publications at least as far back as the 1970s. The first "formal" documentation of kit foxes in Bakersfield was provided by Jensen (1972), who documented the presence of family groups within the city limits. Also, beginning in the mid-1970s, Dr. Ted Murphy at California State University-Bakersfield began conducting work with kit foxes in Bakersfield. This work included placing radio-collars on a handful of foxes, recording observations of foxes from around the city, fielding queries from city residents, attempting to rescue foxes in distress (e.g., trapped in buildings), and conducting rehabilitation work with foxes found injured. Kit foxes and active dens were documented along the Kern River corridor during formal surveys conducted in 1987 (Beedy et al. 1987).

However, foxes may have been present in Bakersfield long before this. The current situations in the west valley towns of Taft and Coalinga may provide models for early occupation of urban environments by kit foxes. Both these towns are relatively small in area (Taft: ~775 ha or 1915 ac; Coalinga: ~505 ha or 1248 ac). Kit foxes are regularly observed in both towns, but based on some limited telemetry

FIGURE 3.9. San Joaquin kit fox in Bakersfield. Photo by Christine Van Horn Job.

and observational data, these foxes use a combination of urban and adjacent exurban environments. If these towns continue to grow in areal extent, they may eventually become sufficiently large such that entire kit fox home ranges will fit within the urbanized areas. This scenario likely occurred in Bakersfield as the city grew. Attracted by abundant food and fewer natural predators, kit foxes may have made incursions into the urban environment, and as a result of their immense adaptability, became full-time urban residents when the city was sufficiently large such that the foxes had enough space to fulfill all their life requirements.

This scenario seems plausible given the advantage of hindsight. However, even into the 1990s, urban kit foxes in Bakersfield were not even remotely viewed as potentially part of a persistent population. To the contrary, they were largely viewed as novelties and remnants of a population in the process of being displaced by urban development. The assumption was that these animals would likely move out into more natural exurban areas, or they would eventually die out by natural or anthropogenic causes; however, this assumption began to be questioned based on certain observations. Recognizable foxes (e.g., missing part of their tail, ear-tagged by Ted Murphy) were observed over multiple years in the same locations, suggesting they were residents in those areas and not being pushed out or dying. Also, litters of pups were commonly observed in urban areas, suggesting that kit foxes were not only living in urban areas, but also successfully reproducing as well.

In the early 1990s, Cypher and Warrick (1993) sampled kit fox scats from and around the CSU-Bakersfield campus. Their analysis revealed that foxes in Bakersfield were not relying solely on garbage, handouts, or pet food left outdoors, but instead their diet appeared to consist largely of natural prey items such as ground squirrels, gophers, birds, and insects. An analysis of previously collected data revealed that urban kit foxes were actually heavier than non-urban foxes, indicating they had access to more food (Cypher and Frost 1999). Then in 1997, a group of biologists initiated a more formal study of the demography and ecology of urban kit foxes. Some of the significant findings of this effort were that kit foxes were more abundant in Bakersfield than had been realized, they were commonly reproducing, and they were using a diversity of areas, including undeveloped lots, parks, school campuses, industrial and commercial areas, golf courses, canal and railroad rights-of-way, and other areas. Foxes exhibited only limited use of residential areas (too many fences, walls, dogs, children, etc.). Some of these findings were presented in a graduate student thesis (Frost 2005). In addition, contrary to the findings of Beedy et al. (1987), kit foxes did not appear to be using the only natural habitat left within the urban area, that being the corridor along the Kern River. Later research revealed that this corridor was frequented by coyotes and bobcats, and their presence likely discouraged use by kit foxes.

Research on urban kit foxes in Bakersfield continued, and in 2000 this effort was incorporated into the Endangered Species Recovery Program (ESRP) administered by California State University–Stanislaus. From 2001 to 2004, the research expanded significantly, with funding from multiple sources (e.g., US Bureau of Reclamation, California Department of Fish and Game, California Department of Transportation, US Fish and Wildlife Service, AERA, Great Valley Center) reflecting growing interest in the urban kit fox population. This research involved intensive monitoring of radio-collared kit foxes to collect data on survival, sources of mortality, reproductive success, home range and movements, den locations, habitat use, food habits, interactions with other species, and interactions with people.

This more intensive research revealed an abundance of fascinating information. The causes of mortality among urban kit foxes were more numerous than for foxes in natural habitats. Whereas predators, primarily coyotes and bobcats, are the primary cause of mortality for non-urban foxes, with occasional vehicle strikes, not too surprisingly vehicles are the primary cause of mortality for urban kit foxes (Cypher 2010). Predators (again, coyotes and bobcats, but also domestic dogs) also caused the deaths of some urban kit foxes, but other causes included toxins (rodenticides in particular), shooting, drowning in fountains, entrapment in irrigation pipes, entombment in dens, and entanglement in soccer and baseball batting cage nets (Figure 3.10).

This last source of mortality is quite curious and a little puzzling. As of this writing, 64 incidents of foxes getting caught in sports nets have been documented, and others likely have gone unreported. In 24 of the incidents, the foxes have died. The foxes surely can see the nets, even at night, and in the case of the soccer and tennis nets, it would be quite easy for the foxes to simply go around them. One possibility is that the foxes do not view the nets as a barrier, as at least in the case of soccer nets, the openings are sufficiently large for a fox to pass through. However, although such passage is easy in rigid structures (e.g., hog-wire fencing), the nets are pliable, and they collapse on the fox as it tries to work its way through the hole, and the fox then becomes entangled. Their panicked struggles result in further entanglement, sometimes to the point of strangulation or injury. Even if foxes are not injured or strangled, they sometimes die of dehydration or exhaustion if they are not discovered for multiple days. This hazard is quite unique to the urban environment.

Despite the more abundant sources of mortality, survival rates (see table 1 in Survival and Mortality Factors) of urban foxes were significantly higher than those of foxes in natural habitats (Cypher 2010). Similarly, reproductive rates were also significantly higher for urban foxes (see table 5 in Reproduction). The higher survival and

A

B

FIGURE 3.10. San Joaquin kit foxes caught in sports nets in Bakersfield. Photos by (a) Christine Van Horn Job and (b) Erica Kelly.

reproduction are likely a function of supra-abundant food coupled with fewer natural predators. Interestingly, although reproductive success is higher (Figure 3.11), litter size is not. Mean litter size is just about 4 pups, with a range from 1 to 9. These values are almost exactly the same as those for non-urban foxes (see table 5 in Reproduction); however, the survival rate of juveniles to age 1 year is among the higher rates recorded (see table 2 in Survival and Mortality Factors). A consequence of these high survival and reproductive rates is that urban kit fox densities are higher than those observed in populations in natural habitat. Estimates of densities in natural lands range from 0.15 to $0.82/km^2$ ($0.06–0.32/mi^2$; Cypher, Deatherage, et al. 2023). The density of kit foxes in Bakersfield was estimated to be about $1.23/km^2$ ($0.47/mi^2$; Cypher, Deatherage, et al. 2023), and this estimate was likely low because of an on-going mange epidemic in this population (see Survival and Mortality Factors). As a result of high survival rates and high density, dispersal potential is limited, which may result in a higher proportion of non-dispersing foxes (Westall et al. 2019), thereby increasing density even further.

In addition to the differences in demographic attributes described above, ecological attributes are also different between urban kit foxes and non-urban foxes. Food is very abundant in the urban environment. As one might expect, foxes are certainly eating considerable quantities of anthropogenic items (Figure 3.12). Foxes find abundant food discarded by people, some dropped as litter and some deposited in refuse bins (e.g., trash cans, dumpsters) where the foxes enter and help themselves. Foxes will also help themselves to pet food left outside, either for family pets kept outdoors or for feral animals, particularly cats. In addition, some people leave food out specifically for foxes. Anthropogenic foods have high digestibility compared with natural foods, and the portion that passes undigested through a fox is generally unrecognizable in standard scat analysis. Occasionally, wrappers are ingested along with foods, and these provide evidence of consumption of anthropogenic items. A dietary analysis based on stable isotopes revealed the amount of anthropogenic food consumed by urban kit foxes is indeed underestimated by traditional scat analysis (Newsome et al. 2010).

Although urban kit foxes indeed consume considerable anthropogenic food, they also consume considerable natural food as well.

A

B

FIGURE 3.11. Litters of San Joaquin kit fox pups in Bakersfield at (a) the East Hills Mall and (b) Sundale Club golf course. Photos by (a) CSUS ESRP and (b) Jason Storlie.

FIGURE 3.12. San Joaquin kit foxes in Bakersfield (a) eating from a pet food bowl, (b) eating discarded food trash, and (c) carrying a captured California ground squirrel. Photos by (a, b) CSUS ESRP and (c) Richard Derevan.

A

B

C

Items commonly found in urban kit fox scats include gophers, California ground squirrels, birds, and invertebrates such as beetles, grasshoppers, cockroaches, earwigs, and moths (Cypher and Warrick 1993; Cypher 2010). Thus, urban kit foxes are not relying solely on anthropogenic foods.

Home ranges of urban kit foxes are smaller than those of nonurban foxes, likely because of the abundance of resources. Space-use is strongly influenced by resource abundance and dispersion, and animals typically scale space-use on distances necessary to fulfill their life history needs. Foxes in the urban environment have home ranges that average 78 ha (193 ac; see table 6 in Space Use, Movements, Dispersal, and Activity) in size, while home ranges of nonurban foxes are generally around 544 ha (1344 ac) in size (Cypher et al. 2013; Cypher, Deatherage, et al. 2023). The amount of space actually used by urban foxes is functionally even smaller than 78 ha (193 ac), because the boundaries of their calculated ranges typically include buildings, fenced or walled areas, and other features inaccessible to the foxes.

In some other urban environments, patches of habitat, largely in an undisturbed condition, have been conserved. Examples include the forest reserve system in Chicago, the matrix of urban and nonurban habitats in the Santa Monica Mountains, and the extensive system of green belts and parks in the northern suburbs of Denver. Many of the wildlife species remaining in these areas rely extensively on these patches, although they also venture into the truly urbanized areas. In Bakersfield, there are virtually no such patches of natural habitat within the urban environment. The one exception is the thin corridor encompassing the Kern River. This corridor was once touted as an area preserved for kit foxes, among other species (e.g., Beedy et al. 1987); however, early in our research, we discovered that we almost never caught kit foxes in this corridor. Furthermore, radio-collared foxes that entered the corridor ended up dead unless they just quickly passed through. The foxes were being killed by coyotes and bobcats. These larger animals indeed rely on patches of natural habitat, and therefore their activities are apparently concentrated in the corridor. Thus, the Kern River corridor in essence constitutes a death trap for kit foxes.

Urban kit foxes in Bakersfield use a variety of urban "habitats" which may be better characterized as land uses. These areas include open space like undeveloped lots, golf courses, power line and railroad corridors, canal banks, parks, and stormwater drainage basins. Areas with more structures and human activity include school and church campuses, industrial and commercial areas, and residential areas of varying density. Kit foxes use all of these, although use of high-density residential areas consisting of single-family homes is limited because of the ubiquitous prevalence of fenced or walled yards, dogs, and concentrated human activity (Frost 2005; Cypher 2010; Deatherage et al. 2021). Areas with higher proportions of campuses and other open space and lower proportions of large roads constitute more optimal areas for urban kit foxes (Deatherage et al. 2021; Cypher, Deatherage, et al. 2023).

As has been emphasized, dens are a critical aspect of kit fox life history requirements. Kit foxes in urban areas appear to have little problem finding locations in which to create dens (Figure 3.13). Essentially, any location with exposed soil that is not frequently disturbed is a potential area for den establishment. Thus, locations like undeveloped lots, power line corridors, railroad rights-of-way, and stormwater drainage basins are all common locations for dens. Kit foxes also establish dens on golf courses, in parks, on school campuses, and in other locations with irrigated lawns. The foxes are apparently undeterred by periodic grass cutting and frequent (e.g., daily) watering. In general, the open spaces listed thus far are not surprising. More surprising is the large number of dens dug in landscape planters and under sidewalks, roads, buildings, and other structures. Essentially, if foxes can reach soil, they can dig a den. However, urban kit foxes have proven extremely resourceful, and not uncommonly they establish non-earthen dens. Such dens include areas underneath portable buildings, seatrains, and other structures, and within pipes and culverts. Consequently, locations for establishing dens are not a limiting factor for urban kit foxes.

Information gathered during two decades of research clearly indicates that the population of urban kit foxes in Bakersfield is robust demographically and ecologically. The population appears to have readily adapted to the urban environment and is generally thriving, absent sarcoptic mange (see Survival and Mortality Factors). Other wildlife species inhabiting urban environments, particularly mam-

A

B

FIGURE 3.13. San Joaquin kit fox dens in Bakersfield: (a) golf course, (b) portable buildings on a school campus, (c) railroad embankment, and (d) a landscape planter. Photos by (a, b) Christine Van Horn Job, (c) Brian Cypher, and (d) CSUS ESRP.

C

D

malian carnivores, may have adapted to and are thriving in these environments, but their presence is typically not without issues or controversy. Examples of urban carnivore issues include coyotes killing pet cats and dogs, coyotes attacking people, even resulting in deaths of small children, striped skunks spraying people and pets, transmission of diseases such as rabies and distemper, raccoons spilling garbage containers, red foxes killing chickens and other domestic birds, and raccoons eating fruit (e.g., Bradley and Altizer 2007; Curtis and Hadidian 2010).

So, an obvious question is whether kit foxes cause any conflict situations when they become abundant in urban environments. The answer seems to be, relatively few. No instances of kit foxes attacking domestic animals have been verified. Indeed, at just 2.3–2.7 kg (5–6 lb) in weight, the reverse is the concern. Kit foxes have been killed by domestic dogs. Their relationship with domestic cats is quite interesting. Cat remains have been found in only a handful of kit fox scats (Cypher and Warrick 1993; ESRP unpublished data), and it is unknown whether these represented kit foxes actually preying on cats (if so, this most likely would be feral kittens) or scavenging on dead cats. Most adult cats outweigh kit foxes; thus, there would actually be considerable risk to any kit fox attempting to prey on a cat. Observations indicate that the two species mostly avoid each other, thereby avoiding the risk of injury. In a study conducted at feral cat feeding stations (Harrison et al. 2011), kit foxes were commonly observed to withdraw when a cat approached the station to feed.

Unlike with urban coyotes, mountain lions, and black bears, there are no documented instances of kit foxes attacking people. Even though they have adapted well to living among humans, kit foxes still tend to avoid and shy away from people. Even kit foxes being fed by people will approach to accept food and then quickly withdraw before consuming it. In Bakersfield, only three instances are known of kit foxes biting people. In all three situations, people were bitten when attempting to assist kit foxes that were entrapped in some way. In two cases, foxes had entered buildings and people were trying to capture them to release them. In the other case, a kit fox was trapped in a soccer net and a student attempted to grab it to help free it. The foxes of course did not recognize that good Samaritans were attempting to help them, but instead they assumed they were being attacked and reacted by defending themselves.

Other species can be substantial nuisances in urban environments. Raccoons and bears tip over trash cans. Deer and rabbits raid vegetable gardens and damage landscape plantings. Squirrels chew their way into roofs and attics. Canada geese deposit copious quantities of fecal material on sidewalks, park lawns, and golf courses. Skunks and pocket gophers damage lawns. However, kit foxes are not known to cause damage. We once received a call from someone complaining that a kit fox was defecating on their walkway each night. Having our suspicions, we investigated and found that the fecal deposits were from western toads, which are abundant in Bakersfield and have scats approximately the same size as those of a kit fox. (However, toad scats characteristically consist exclusively of finely ground invertebrates.)

Kit foxes in urban environments seem associated primarily with two nuisance issues. One is that they have been implicated in promoting flea infestations in portable classrooms, which they commonly use as denning structures. The actual role of kit foxes in such infestations is unclear. Although kit foxes undeniably carry fleas, in almost all situations where infestations have been reported, other species such as cats and skunks have also been found residing under the portable buildings. In these situations, it is unknown whether one species is more responsible for the fleas or whether all are equal contributors. Regardless, the situation is easily resolved by treating the inside of the buildings to kill fleas or by excluding animals from gaining entry underneath the portables.

The other nuisance situation involves kit foxes digging dens in inconvenient locations (Figure 3.14). Such locations have included golf-course sand traps, landscape beds, locations that receive substantial foot traffic (therefore creating a risk of injury from stepping into a den entrance), and construction sites. On the construction sites, the foxes are apparently attracted by the soft, disturbed soil that is easy to dig in. (Also, the foxes may have had a den in the area that was destroyed by the construction and are attempting to construct a new den.) In most situations, the foxes eventually move on to other dens, and the offending den can be excavated and collapsed. Where foxes are more persistent or a den needs to be closed more quickly, one-way doors have been used, which allow foxes to exit and prevent re-entry. Once foxes (or other species) are no longer using the dens (as determined by automated cameras, tracking medium, or other means), the den can be excavated and collapsed. A unique situation involved a den constructed in soil right alongside a building. The soil excavated by the foxes had piled up in front of the back door into the building, thereby rendering the door inoperable. The employees of the business in that building repeatedly cleared away the soil, but periodically, the foxes would excavate additional soil, thereby re-creating the problem. The foxes did eventually move on, thus ending the problem. More problematic are situations involving natal dens. Foxes occupy these dens for longer periods of time, and by law, the foxes and dens cannot be disturbed. Thus, the area simply has to be avoided until the foxes eventually move on, and this can result in a nuisance situation lasting for a number of weeks.

The San Joaquin kit fox population in Bakersfield essentially constitutes a bonus (i.e., unexpected) population from a conservation perspective. A persistent population, and particularly one that is thriving for the most part, was not anticipated in early conservation and recovery planning. However, this population (at least as of 2022) is one of the largest of the remaining populations (see Distribution and Habitat Preferences) and may even be larger than or at least similar in size to the Panoche Valley core population identified in the recovery plan for kit foxes (USFWS 1998). Furthermore, the Bakersfield kit fox population may be the only one that is actually growing! All other remaining populations are either stable or decreasing. For some populations, such as those on the Carrizo Plain and Panoche Valley, kit foxes have already occupied all the available suitable habitat, and thus there is no further room for population expansion. For other populations, the available habitat is still decreasing because of continuing conversions to agricultural crops, industrial operations (e.g., oil and gas production, solar farms), mineral extraction, and other uses. In what seemingly constitutes a paradoxical situation, the urban environment of Bakersfield is expanding at a rapid pace, and kit foxes are rapidly colonizing the new urban areas. The population growth is occurring because the lands that are being developed are mostly agricultural lands in which kit foxes were not

A

B

FIGURE 3.14. San Joaquin kit fox dens in inconvenient locations: (left) a golf course sand trap and (right) an active construction site. Photos by (a) Brian Cypher and (b) Christine Van Horn Job.

present following their conversion to cropland. Thus, the Bakersfield kit fox population is increasing and may continue to do so as the city is projected to continue growing (World Population Review 2021).

The Bakersfield population can potentially contribute significantly to conservation and recovery efforts for San Joaquin kit foxes (Cypher 2010; Cypher and Van Horn Job 2012; Cypher, Deatherage, et al. 2023). As stated above, it is a bonus population that results in a greater overall number of kit foxes within the historic range. Likewise, genetic diversity of the overall population is enhanced by the additional individuals and extra location. Indeed, Wilbert (2013) found that the Bakersfield kit fox population was genetically robust (i.e., no inbreeding, genetically diverse) and harbored unique alleles not found in other populations. The presence of this large, genetically diverse population enhances the long-term viability of the overall San Joaquin kit fox metapopulation. Furthermore, this population potentially provides other important benefits (Bremner-Harrison and Cypher 2007; Cypher 2010; Cypher and Van Horn Job 2012). Because it is demographically robust with high reproductive rates, this population could serve as a source of individuals for reintroductions into any areas with suitable habitat but where kit foxes are not present (e.g., areas where they might have been extirpated by some past event such as disease or an anthropogenic cause that has since been mitigated, or restored habitat on retired agricultural lands).

Finally, the presence of kit foxes in close proximity to humans creates valuable outreach opportunities. During our many years working with the Bakersfield kit fox population, we found that people like seeing the kit foxes, take an interest in them, and are concerned about them. This was abundantly evident in the number of times we were challenged by individuals concerned that we were trapping foxes to remove them or harm them in some way. We had to assure those concerned citizens that we were trapping to help the foxes and that they would be released unharmed at the capture site. It was also evident in the many calls, text messages, and emails that we received from people over the years informing us of injured kit foxes, dead foxes, or foxes potentially in harm's way, or just providing information about foxes or fox dens they had observed. Additionally, we conducted a small survey effort in 2004 which confirmed that many people were interested and concerned about the kit foxes in Bakersfield (Bjurlin and Cypher 2005). Interestingly, the survey also revealed that people who had observed foxes in Bakersfield not only favored conserving the urban foxes but were also more likely to favor conservation of San Joaquin kit foxes in natural lands and throughout their range. In essence, the Bakersfield kit foxes function as excellent ambassadors for their species.

The Bakersfield population is remarkably robust demographically and ecologically. In a sense, it may have been a victim of its own success in the form of a sarcoptic mange epidemic (see Survival and Mortality Factors). Nevertheless, it provides evidence of the potential for kit foxes to adapt to and thrive in urban environments, with the ultimate result being additional populations in areas where kit foxes were expected to be extirpated. As of 2021, small numbers of kit foxes were also present in the towns of Taft (pop. ~10,000) and Coalinga (pop. ~18,000), both located on the west side of the San Joaquin Valley. At least some of these foxes appear to be living primarily in the urban environment while others are using a mix of urban and nearby natural lands. Both towns are bounded by natural habitat, similar to Bakersfield, and this likely facilitates use of these urban areas by kit foxes.

An interesting situation to watch going forward will be whether other similarly robust urban kit fox populations eventually appear. Other sizeable towns within the historic range of the San Joaquin kit fox include Delano, Hanford, Tulare, Porterville, Visalia, Fresno, Los Banos, and Patterson. However, these towns are surrounded by agriculture or low suitability habitat, and therefore no foxes are present nearby to colonize or use these urban areas. Kit foxes may have ventured into these towns when they still had connectivity to suitable habitat, but connectivity may have been severed before the urban environments of these towns were sufficiently large to sustain a persistent kit fox population. An idea worthy of consideration is whether kit foxes could, or even should, be intentionally introduced into certain urban areas in an effort to establish additional populations. Bakersfield could potentially serve as a source population, as this population is demographically robust and could easily sustain the removal of some foxes. An additional advantage is that foxes from Bakersfield would already be urban-adapted. In all reality, any new urban populations would likely need to be treated as experimental (a designation in which a population of endangered species is afforded fewer protections) as regulatory relief might be necessary to reduce opposition over fears of consequences due to accidental deaths of kit foxes. However, the benefits to recovery of the San Joaquin kit fox could be considerable.

KIT FOXES AND CLIMATE CHANGE

Climate change has been a prevalent and growing topic in ecology and conservation for several decades now. Indeed, numerous entire professional careers have been built solely upon modeling climate change, modeling ecological effects, estimating impacts to species and ecosystems, and developing conservation strategies to address these impacts. Such efforts have extended to the San Joaquin Valley ecoregion in general, and to the San Joaquin kit fox in particular. Although climate change modeling has become incredibly sophisticated, it is impossible to account for all factors, particularly possible unforeseen future ones. Consequently, there is a high level of uncertainty and an array of potential scenarios. Although it seems clear that much of the world will become warmer, less certain is whether particular regions will become wetter or drier. This uncertainty has considerable implications for kit foxes (and obviously many other species), and the good news is that kit foxes may actually fare quite well under some of the potential scenarios.

The diversity of potential scenarios results from two primary processes: ecological changes, and socioeconomic impacts caused by climate change. Changing temperature regimes and precipitation patterns could produce a host of ecological effects. These changing patterns would very likely affect the composition and structure of vegetation communities, and these altered habitat conditions would subsequently affect animal communities. The relative abundance of species in a given region could change, with some species disappearing and new species potentially becoming established. Wetter or cooler conditions, or both, could increase the density of ground cover and shrubs, which would reduce habitat suitability for kit foxes as well as for their primary prey, kangaroo rats. Conversely, warmer and drier conditions would have the opposite effect by reducing ground cover and shrub density, thereby potentially increasing suitability for kit foxes and kangaroo rats. Changing temperature and precipitation regimes could also directly affect the presence and abundance of certain animal species, particularly any that might have narrow thermal and moisture tolerances. Kit foxes as a species have fairly wide tolerances, as evidenced by their occurrence in extremely hot regions such as Death Valley as well as in fairly cold regions like northern Utah, southern Oregon, and southern Idaho. Annual precipitation within their range varies from <10 cm (4 in) to ~125 cm (49 in) (WRCC 2021).

Climate change effects would affect kit fox distribution and abundance at two scales. Within the current range of the San Joaquin kit fox, these effects could cause some areas to become more suitable and some areas to become less suitable, which would affect local abundance of kit foxes. Perhaps more significantly, climate change effects could produce an expansion or contraction of the range of the San Joaquin kit fox (USFWS 2020b). Overall cooler or wetter regional conditions would result in less habitat being suitable for this species. However, warmer or drier regional conditions could result in habitat outside the range becoming more suitable, and thus kit foxes could potentially expand into newly suitable areas, thereby expanding their range.

An analysis of climate change effects on the San Joaquin kit fox was conducted by Entrix, Inc. (2010). This analysis indicated that under a leading climate change projection model for the San Joaquin Valley, suitable habitat for the San Joaquin kit fox could expand substantially, particularly west into the Coast Ranges and north into the northern San Joaquin Valley and the southern Sacramento Valley (Figure 3.15). Another analysis, conducted by the USFWS (2020b), considered two plausible climate change scenarios based on recent models. The range of the San Joaquin kit fox would contract somewhat under one scenario (warmer and wetter conditions), and it would expand under the second scenario (hotter and drier conditions). Thus, the impact of climate change on the San Joaquin kit fox is characterized by considerable uncertainty, but it is encouraging that under some plausible scenarios, kit foxes may do just fine.

San Joaquin kit foxes could also be affected by socioeconomic impacts resulting from climate change. The availability of water for ag-

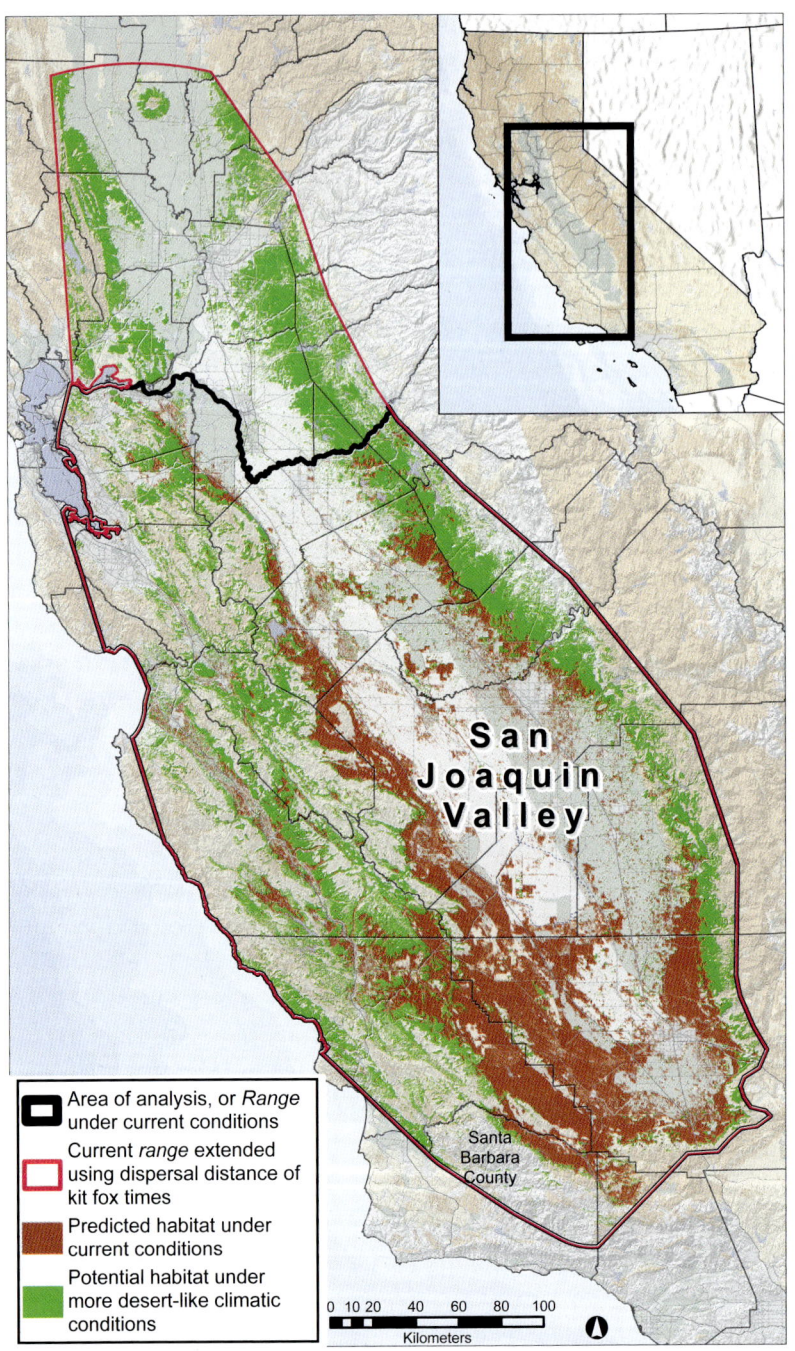

FIGURE 3.15. Potential range expansion by San Joaquin kit foxes under plausible modeled climate change scenarios. Prepared by Scott Phillips.

Legend (in image):

- Area of analysis, or *Range* under current conditions
- Current *range* extended using dispersal distance of kit fox times
- Predicted habitat under current conditions
- Potential habitat under more desert-like climatic conditions

San Joaquin Valley

Santa Barbara County

0 10 20 40 60 80 100 Kilometers

ricultural and urban uses is an immense issue in the San Joaquin Valley. The primary sources of water in this region are spring runoff from the winter snowpack in the Sierra Nevada, water from northern California and the Delta that is funneled south via the California Aqueduct, and groundwater pumped from subterranean aquifers. The first two sources vary annually in precipitation, and all available water from these sources is maximally allocated to various users, particularly agriculture and urban areas. Groundwater in the San Joaquin is primarily in a state of overdraft. In most areas of the valley and in most years, the amount of water being pumped exceeds the rate of recharge, which led to the passage of the Sustainable Groundwater Management Act (SGMA) of 2014 (see Kit Foxes and Agricultural Lands). Thus, water availability is already limited in the San Joaquin Valley and will become even more limited as urban populations increase in this rapidly growing region and as measures such as SGMA are implemented.

Clearly, any changes in water availability associated with climate change will have significant consequences. Forecasts are that hundreds of thousands of acres will likely be taken out of agricultural production as a result of SGMA alone (USFWS 2020b; Butterfield et al. 2021). Reductions in water availability due to climate change could result in even more land being taken out of production (i.e., retired). Some proportion of these lands could revert naturally or be restored to habitat suitable for kit foxes and other associated species. Thus, socioeconomic impacts from climate change could result in significantly more habitat becoming available within the current range of the San Joaquin kit fox. Conversely, it is also possible under cooler/wetter scenarios that additional habitat could be converted to agricultural or urban uses, resulting in even less habitat for kit foxes and other species (USFWS 2020b). The uncertainty associated with climate change is immense and will likely continue to be so for many decades.

Chapter 4

CONSERVATION

This chapter is both retrospective and prospective. It begins with summaries of past research and conservation efforts on San Joaquin kit foxes that include nods to the individuals and organizations that have contributed significantly to these efforts. Consistent with a main objective of this book, the intent of these summaries is to document what has been accomplished so as to help identify future needs. Before listing and discussing these needs, this chapter considers politics and attitudes regarding San Joaquin kit foxes, as they significantly influence the course, invested resources, and ultimate success of conservation efforts. Thus, a look back and then a look forward.

HISTORY OF RESEARCH EFFORTS

The San Joaquin kit fox was first recognized as a distinct subspecies of kit fox by Grinnell and colleagues in their 1937 publication "Fur-bearing Mammals of California" (Grinnell et al. 1937). In this book, they summarized data collected over the previous two decades on kit foxes and other carnivores in the state. As was typical of biologists of that period, especially those like Grinnell and colleagues associated with a museum (in this case, the UC-Berkeley Museum of Vertebrate Zoology), they gathered data primarily from specimens collected for inclusion in the museum. They examined 21specimens of San Joaquin kit foxes collected from 23 June 1918 to 12 August 1929, and the data recorded were primarily morphological attributes. However, during their field work, Grinnell and his colleagues also meticulously recorded details on habitats, behavior, and associations with other species. This information constituted the initial life history description for this species. In addition to providing details on habitat associations, the observers also reported the strong relationship between kangaroo rats and kit foxes. They noted that kit foxes were most abundant in areas where kangaroo rats were abundant.

After this initial work by Grinnell and colleagues, little formal research was conducted on San Joaquin kit foxes for the next several decades. However, enough information was gathered to recognize that the San Joaquin kit fox was rapidly declining in number such that it was placed on the very first formal list of imperiled species in the United States in 1967 (USFWS 1998). In response to this, a series of extensive surveys were conducted over the next couple of decades, primarily by flying transects through large portions of the range of the San Joaquin kit fox and counting dens from the air. The surveys were conducted by Lyndal Laughrin (1970), Charles Jensen (1972), and Stephen Morrell (1975). Estimates of kit fox numbers were derived based on these surveys, but the estimates were clearly fraught with errors and issues. For example, in most of these surveys, it was assumed that at least one kit fox was associated with each den. Sufficient research had not been conducted at this point to reveal that each kit fox pair or family group has multiple dens scattered throughout its home range (sometimes 20 or more; see Dens). Thus, the numbers of foxes derived from these surveys were generally markedly inflated estimates. Also, it is unknown to what extent surveyors were able or even attempted to distinguish California ground squirrel burrows, badger digs, and other excavations from actual kit fox dens.

Quantitative ecological studies on San Joaquin kit foxes were not really conducted until the 1970s. Data collection was immensely facilitated by live-trapping and marking kit foxes and tracking them using radio collars. Stephen Morrell (1972, 1975) studied kit foxes in the Buena Vista Valley in Kern County using radio telemetry and provided some of the first quantitative details on the demography and ecology of kit foxes. Donna Knapp (1978) conducted a landmark study in Kern County where she quantified the impacts to kit foxes from the conversion of habitat to agricultural uses. Not surprisingly, she found that some kit foxes were entombed during disking activities, and foxes that escaped entombment were displaced.

Five significant research efforts were initiated in the 1980s. The California Department of Fish and Game initiated quarterly spotlight surveys for kit foxes along several routes in the San Joaquin Valley and nearby Carrizo Plain. The two routes in the Carrizo Plain region were still being surveyed as of 2022. These surveys were valuable in providing multi-decade indices of kit fox population trends. Another effort was carried out by Ted Murphy, a professor at the California State University-Bakersfield, who began collecting information on kit foxes inhabiting the city of Bakersfield. He even placed radio collars on a couple of these urban foxes. A third effort was initiated in the late 1980s by Linda Spiegel and a research team with the California Energy Commission (CEC). The primary objective of this work was to investigate the effects of oil and gas production on kit foxes in western Kern County. They examined a variety of demographic and ecological attributes (Spiegel 1996) and found that kit foxes persisted in areas with a high density of wells and other facilities as long as the area still supported patches of natural habitat. A fourth effort initiated in the late 1980s was ecological studies of kit foxes on the Carrizo Plain conducted by Katherine Ralls and P. J. White of the Smithsonian Institution. This work was conducted primarily during 1988–1992 and, as with the CEC study, examined a number of demographic and ecological attributes of kit foxes. A series of journal articles resulted from this work, and some of the main findings included identification of coyotes as a main predator on kit foxes, the importance of dens to kit foxes, response of kit foxes to variation in prey availability, and resource competition between coyotes and kit foxes.

Arguably the most significant effort initiated in the 1980s was that of EG&G Energy Measurements, Inc., a contractor conducting environmental work for the US Department of Energy. This work really started in 1979 and continued under EG&G until 1995, and then it continued until 1998 under subsequent contractors that absorbed EG&G staff. The research was initially directed by Thomas O'Farrell and then later by Thomas Kato and Brian Cypher. Most of the work conducted by EG&G on San Joaquin kit foxes was at the Naval Petroleum Reserves in California, consisting of two massive oil fields,

NPR-1 (or Elk Hills) and NPR-2 (or Buena Vista Hills). Chevron also owned a portion of NPR-1 and was a contributing sponsor of the research conducted there. The primary emphasis of the research was to assess the effects of oil and gas production activities on the abundance and distribution of kit foxes, similar to the CEC effort, as well as other federally or state listed species or species of conservation concern. This work resulted in the preparation of dozens of technical reports and journal articles. Much of the main demographic and ecological information on kit foxes was summarized in a monograph (Cypher et al. 2000) covering the most intensive period of investigation (1980–1995). Among the main findings was that other factors such as terrain, predators, and food abundance and dispersion had much more significant effects on kit fox abundance and distribution than did oil and gas production activities. In particular, higher predation risk was associated with more rugged terrain, and kit fox abundance and reproduction varied with food availability, particularly kangaroo rats, which in turn was strongly influenced by annual precipitation and corresponding primary productivity.

EG&G also worked on San Joaquin kit foxes on sites other than the Naval Petroleum Reserves. In the 1980s, EG&G staff conducted surveys for kit foxes and other rare species at multiple locations in the San Joaquin Valley, particularly on government lands (e.g., US Bureau of Land Management) and some military lands (e.g., Naval Air Station-Lemoore). The purpose of most of this work was simply to determine whether kit foxes were present on these lands. Under contract with the US Army, EG&G also conducted demographic and ecological studies on kit foxes at the Camp Roberts Army National Guard Training Site and the Fort Hunter Liggett, both located in the Salinas Valley of San Luis Obispo and Monterey Counties on the margin of the range of San Joaquin kit foxes. Most of this work was conducted in the late 1980s and early 1990s, and eventually it was presented in a series of technical reports and some journal articles. A small amount of research, mostly population trend monitoring, continued at Camp Roberts in the late 1990s when the California Polytechnic State University at San Luis Obispo assumed the lead on the project in 1995.

In the 1980s and 1990s, several other smaller but important research efforts were conducted. These included an assessment by Antonio Balestreri (1981) of the status, distribution, and ecology of kit foxes at Camp Roberts and in the Salinas Valley; a study by Frank Hall (1983) of kit foxes at a site at the very northern margin of their range in Alameda County; a study by Laurie Briden and others (1992) on the ecology of kit foxes in a small satellite population in western Merced County; and taxonomic genetic work by Alan Mercure and others (1993) under Robert Wayne at the University of California-Los Angeles. Beginning in the 1990s, Katherine Ralls and collaborators have conducted a number of research projects on San Joaquin kit fox genetics and social ecology.

Another significant body of research on San Joaquin kit foxes has been conducted by the California State University-Stanislaus (CSUS), Endangered Species Recovery Program (ESRP). ESRP was founded by Daniel Williams at CSUS in 1992, and he coordinated the kit fox work in the 1990s, assisted by Patrick Kelly. In the 1990s, the work consisted of surveys for kit foxes and also investigations of use of agricultural lands by kit foxes (Warrick et al. 2007) and movements of foxes in the Panoche Valley region (ESRP, unpublished data). Brian Cypher, who began working with kit foxes with EG&G in 1990 and led the work at the Naval Petroleum Reserves in the late 1990s, joined ESRP in 2000 and coordinated much of the kit fox work conducted by ESRP after that. The investigations have been extensive and include studies on road effects, solar farm effects, use of road crossing structures, use of artificial dens, demographic and ecological attributes of populations in core and satellite areas, interactions with coyotes and other predators, habitat suitability, demography and ecology of kit foxes in urban environments, social ecology, effects of sarcoptic mange, genetic attributes and gene flow, and conservation and recovery planning. Results of this work have been reported in numerous technical reports and journal articles. Significant collaborators include the California Department of Fish and Wildlife, University of California-Davis, CSU-Bakersfield, CSU-Fresno, Center for Natural Lands Management, California Living Museum, and private consulting firms, including Althouse & Meade, Inc., H.T. Harvey and Associates, and McCormick Biological, Inc.

All the research efforts listed above have contributed to a greater understanding of the biology and ecology of San Joaquin kit foxes, their distribution and habitat preferences, conservation status, and conservation strategies. As with any situation, the more information available, the better informed any decisions on conservation and recovery strategies. Rarely, if ever, is there a situation where we have all the information we would ideally like to have. The same definitely applies to San Joaquin kit foxes. Thus, there are still some critical outstanding information needs (see Research and Conservation Needs).

HISTORY OF CONSERVATION EFFORTS

Conservation and the San Joaquin kit fox are essentially inextricably linked. Almost immediately upon recognizing the San Joaquin kit fox as a distinct subspecies of kit fox, researchers also recognized that it was rapidly declining in abundance. In their landmark book on fur-bearing mammals in California, Grinnell and colleagues (1937) expressed concern that San Joaquin kit fox habitat was disappearing at an alarming rate and that kit foxes were also dying in large numbers as a result of harvesting and also predator control programs. The predator control programs employed various toxicants, particularly sodium cyanide, strychnine, and sodium monofluoroacetate (Compound 1080) that killed animals rather indiscriminately. Thus, even though coyotes were typically the primary target, many other species, including kit foxes, were also killed. The concept of conserving wildlife was just emerging in the early twentieth century, so there were not immediate calls for conserving San Joaquin kit foxes, but there was recognition that the population was on a downward trajectory.

Population estimates are difficult to obtain for most carnivores, and especially species like kit foxes that are small, secretive, and spend each day hidden in dens. Based on estimated densities of 0.39–0.54/km^2 (1–1.4/mi^2; Grinnell et al. 1937; Morrell 1975) and an estimated 22,447 km^2 (8667 mi^2) of suitable habitat, the pre-1930 population was estimated to be 8667–12,134 foxes (USFWS 1983), although no indication was given as to whether this included just breeding adults or all foxes. The higher number in particular may be an overestimate. as it is based on an assumption of two kit foxes per active den observed, and each kit fox clearly uses more than one den (see Dens). In the late 1960s, Laughrin (1970) estimated that range-wide density was 0.5 foxes/km² (0.2/mi²) and estimated the total population at 1000–3000. A few years later, Morrell (1975) estimated the number of foxes to be 14,831 based on the area of valley floor and foothill lands in 14 counties in which observations of kit foxes had been documented. This estimate was adjusted to 6961 foxes based on habitat loss documented in each of those counties (USFWS 1983). However, even this adjusted estimate was likely overly optimistic as fox populations were never confirmed in a number of the counties (e.g., Alameda, Contra Costa, San Joaquin, Santa Clara, Stanislaus), and the estimate was again based on an assumption of two foxes per active den observed in aerial surveys.

Cypher et al. (2013) estimated the potential high-quality and medium-quality habitat remaining in 20 counties at 9835 km² (3797 mi²). They then divided this total by the average home range size for kit foxes in high-quality habitat and then assumed two adult foxes per home range to produce an estimate of 3616 foxes. However, this estimate was considered overly optimistic as not all counties had confirmed kit fox populations, some of the habitat was in patches too small to support even one kit fox home range, and kit fox home ranges in medium-quality habitat would be larger (see Space Use, Movements, Dispersal, and Activity), meaning there would be even fewer foxes. Also, no persistent kit fox populations were known to occur in medium-quality habitat at the time. A more realistic estimate is likely obtained by taking the total high-quality habitat in the eight counties known to have kit fox populations: Kern, San Luis Obispo, Kings, Fresno, Tulare, Merced, Santa Barbara, and San Benito. This estimate of 4069 km² (1571 mi²) would support 748 home ranges and 1496 breeding adult kit foxes. This estimate may be a bit high as some of the habitat is in patches too small to support even one kit fox home range, and because of mortalities, two adults may not be present in all home ranges at a given time. However, the number is likely an underestimate of the total foxes in that it includes only breeding adults and does not include helpers or other non-dispersing young, dispersing foxes that have no established home range, and young of the year. The estimate also does not include foxes occurring in urban areas, potentially a substantial number (see Kit Foxes and Urban Areas). Regardless, the range-wide population of the San Joaquin kit fox has declined significantly, and the number of remaining foxes is modest at best.

Regulatory Conservation Actions

With the recognition that San Joaquin kit fox numbers had declined and were continuing to decline, formal regulatory actions were initiated to conserve the species. In 1965, the San Joaquin kit fox was designated a "protected fur-bearer" by the California Fish and Game Commission. On October 15, 1966, the Endangered Species Preservation Act became law (80 Stat. 926; 16 U.S.C. 668aa(c)). Section 1(c) of this act directed the Secretary of the Interior to consult with "the States, interested organizations, and individual scientists" and prepare a list of native fish and wildlife that are threatened with extinction. Stewart Udall was the Secretary of the Interior at the time and acted as directed. A list of 78 species (14 mammals, 36 birds, 6 reptiles and amphibians, and 22 fish) was published in the Federal Register on March 11, 1967 (USFWS 1967). The San Joaquin kit fox was one of the mammal species on the list. It is unknown what authorities or individuals were consulted to prepare this list, but clearly they felt strongly that the San Joaquin kit fox should be included. Thus, clearly there was recognition that this subspecies of kit fox was declining and needed protection. (Interestingly, a co-occurring species, the blunt-nosed leopard lizard [*Crotaphytus wislizenii silus*; now *Gambelia sila*] was also on this initial list, and like the San Joaquin kit fox, is still on the Endangered species list.) In 1971, the San Joaquin kit fox was listed as Threatened under the new California Endangered Species Act. As of 2023, the San Joaquin kit fox was still listed under both the federal and state acts.

Designation as a protected fur-bearer and listing under the federal and state endangered species acts made it illegal to harvest or purposely kill kit foxes. Interestingly, the protected fur-bearer status also extended to desert kit foxes, likely because subspecies of kit foxes in California cannot be distinguished (other than genetically). Thus, harvests of San Joaquin kit foxes and all other kit foxes ended. In addition, predator control programs were declining in the San

Joaquin Valley or transitioning to more individual-specific as well as nonlethal methods. Listing also imposed restrictions on such programs within the range of the San Joaquin kit fox, including prohibitions on use of toxicants (e.g., M-44 devices) and burrow fumigants, and prohibitions on use of Conibear-type traps, snares, and deadfall traps, and limiting use of leg-hold traps to those with padded jaws (USDA 1997). Thus, some major sources of anthropogenic mortality were eliminated, but habitat loss continued at a rapid pace.

In 1983, a recovery plan was completed for the San Joaquin kit fox (USFWS 1983). However, very little was known at the time about kit fox ecology, demography, or population status. Thus, the plan called primarily for mitigating mortality factors and conserving about 35,000 acres of additional habitat. This plan would soon be deemed inadequate.

In the 1970s and 1980s, there was a growing recognition that other species that shared the same habitat with kit foxes were also becoming disconcertingly rare. These species included the giant kangaroo rat, blunt-nosed leopard lizard, Tipton kangaroo rat, Fresno kangaroo rat, San Joaquin antelope squirrel, and even plants such as the Bakersfield cactus, California jewelflower, San Joaquin woolly-threads, Kern mallow, and Hoover's woolly-star. All these species were eventually listed as Endangered or Threatened under the federal Endangered Species Act or California Endangered Species Act, or both (USFWS 1998). The San Joaquin kit fox and these other species all occur in arid saltbush scrub, jointfir scrub, or non-native grasslands. Thus, the USFWS initiated work on a multi-species plan. In 1998, the USFWS issued a new recovery plan (Figure 4.1), Recovery Plan for Upland Species of the San Joaquin Valley, California (USFWS 1998). In addition to the San Joaquin kit fox, the plan included 33 other animals and plants occurring in the San Joaquin Valley ecoregion. Some of these species were federally or state listed, as mentioned previously; however, the bulk of the species in the plan were not listed but instead were "species of conservation concern." Many of these species overlapped in distribution with kit foxes. Because the kit fox had the largest distribution and also because each fox requires large areas of habitat compared with individuals of the other species, the San Joaquin kit fox was designated an "umbrella

FIGURE 4.1. Cover of the Recovery Plan for Upland Species of the San Joaquin Valley, California. This plan includes conservation measures for the San Joaquin kit fox. Photo by Brian Cypher.

species" for conservation. The assumption was that if enough habitat was conserved to support kit foxes, the many co-occurring species would benefit as well.

The 1998 plan has served as the primary guidance and road map for kit fox conservation. Over time, however, the plan has become increasingly outdated as recovery tasks have been completed and new conservation needs identified. Some of this progress has been captured in five-year reviews completed in 2010 and 2020 (USFWS 2010, 2020a) and also in newly implemented "species status assessments" (USFWS 2020b). The USFWS has not had the staff, time, or financial resources to revise the 1998 plan, and thus the five-year reviews and species status assessments are increasingly becoming the primary guidance documents for conservation and recovery. However, the plan, the reviews, and the assessments all emphasize that habitat loss is the primary threat to San Joaquin kit foxes and the other species and that habitat conservation is the primary action needed to conserve and recover these species.

Habitat Protection and Other Conservation Efforts

Little appears to have occurred with regard to conservation efforts for the San Joaquin kit fox until the 1960s. The first conservation efforts for San Joaquin kit foxes were essentially the same as the early research efforts. In the 1970s, surveys were conducted throughout the range to determine the distribution and abundance of the species (see History of Research Efforts), and to assess the amount of remaining habitat. In the late 1980s, the California Energy Commission conducted an assessment of habitat conditions on 21,108 km^2 (8150 mi^2) in the southern San Joaquin Valley (Anderson et al. 1991). They determined that on the 14,763 km^2 (5700 mi^2) they were able to access and evaluate, only 3.6% was in a good natural condition. Of the 7640 km^2 (2950 mi^2) of valley floor habitat (where kit foxes and associated rare species are primarily found), only 2.9% was in good condition. They concluded that their results clearly illustrate the significant loss of natural habitat in the southern San Joaquin Valley. Many smaller-scale surveys and inventories were completed as well around this time.

Thus, efforts began increasing in the 1980s to conserve lands with arid shrubland and grassland habitats to benefit San Joaquin kit foxes and associated species. One of the larger habitat acquisitions occurred in 1988 when the US Bureau of Land Management (BLM), California Department of Fish and Game (now California Department of Fish and Wildlife, CDFW), and The Nature Conservancy collaborated to purchase 33,184 ha (82,000 ac) of land in eastern San Luis Obispo County (Alagona 2013), which in combination with other BLM lands became the Carrizo Plain Natural Area (Figure 4.2). Additional lands have been added and the area now is about 80,900 ha (200,000 ac) in size. This area was designated as the Carrizo Plain National Monument by Presidential Executive Order issued by Bill Clinton on January 17, 2001, using his authority under Section 2 of the Antiquities Act (USBLM 2010).

Additionally, by 1994, more than 64,750 ha (160,000 ac) of upland habitat had been conserved through land purchases and transfers (Alagona 2013). Many of these lands came to the CDFW as Ecological Reserves (John Battistoni, CDFW, personal communication). A very significant advance in habitat conservation began in 1994 with

FIGURE 4.2. San Joaquin kit fox at the Carrizo Plain National Monument. Photo by Brian Cypher.

the approval and implementation of the Metropolitan Bakersfield Habitat Conservation Plan (MBHCP). This was one of the first HCPs implemented in the United States, and at the time of its implementation it covered the largest area (1057 km² or 408 mi²) of any HCP. In brief, the plan permitted habitat destruction within the implementation area, and in return, habitat would be conserved elsewhere and endowments for management would be provided (Metropolitan Bakersfield Habitat Conservation Plan 1994). As of 2021, the MBHCP has resulted in the conservation of more than 9375 ha (24,000 ac) of habitat (J. Battistoni, CDFW, personal communication). By 2009, the San Joaquin kit fox had been included in 20 additional HCPs throughout the San Joaquin Valley region (Alagona 2013).

In the two decades following the release of the 1998 San Joaquin Valley Upland Species Recovery Plan, considerably more habitat has been conserved through a variety of processes. Substantial habitat conservation plans have been approved, such as those for the Elk Hills oil field, ARCO/AERA oil fields, and Kern County landfills, with additional HCPs also under consideration. Significant acreage of good quality kit fox habitat has also been conserved as mitigation for new solar farms. Notable projects include more than 12,140 ha (30,000 ac) conserved in association with the Topaz Solar Farms and the California Valley Solar Ranch in the northern Carrizo Plain regions, more than 10,500 ha (26,000 ac) associated with the Panoche Solar Project in the Ciervo-Panoche region, more than 970 ha (2400 ac) associated with the Wright Solar Farm in western Merced County, more than 200 ha (500 ac) associated with the Blackwells Corner Solar Farm in northwestern Kern County, and more than 1800 ha (4455 ac) associated with the California Flats Solar Farm in northeastern San Luis Obispo County (Cypher, Boroski, et al. 2021). Additionally, about 1370 ha (3500 ac) of habitat has been conserved in association with the California High Speed Rail Project (J. Battistoni, CDFW, personal communication). Also, a number of mitigation banks have been established, and conservation easements have been purchased on private lands.

In addition to the habitat conservation efforts, considerable research has been conducted on kit fox ecology, mortality factors, and response to land use changes. Some of the more notable efforts include kit fox ecology in urban areas, kit fox response to solar farms, epidemiological studies especially regarding sarcoptic mange and distemper, and basic demography and ecology studies in various population areas in an effort to capture demographic and ecological variation throughout the range of the species.

As of this writing, considerable conservation efforts have been expended on behalf of the San Joaquin kit fox, including habitat conservation as described above, more extensive implementation of impact mitigation measures, and research. Unfortunately, the species seems little closer to recovery than it was in 1967. There are several reasons for this. First, although conservation efforts have resulted in the protection of substantial acres of habitat, the total available habitat has declined significantly since 1967 because of ongoing conversion to other uses, primarily agriculture and urbanization. Thus, there is less overall habitat and therefore probably fewer foxes than there were in 1967. A second factor is that the recovery criteria in the 1998 plan (USFWS 1998) are sometimes a bit vague. They essentially call for a certain proportion of habitat to be conserved in a number of population areas and for the fox population in each of these areas to be stable or increasing; however, the boundaries of those areas were never explicitly defined. The USFWS did not want hard lines drawn on maps for fear that landowners whose lands occurred within those lines would strenuously object and obstruct recovery efforts. Thus, it is impossible to know whether or when the designated target for a given area (e.g., 90% of Lokern) has been achieved. A third and critical factor is that even if the habitat conservation targets in the recovery plan were achieved, it is unknown whether this would be sufficient to maintain persistent, self-sustaining populations. This can never be known with complete confidence, but a highly robust estimation can be produced through population viability analysis. Indeed, the 1998 plan acknowledges this and recommends that metapopulation viability analyses be completed for San Joaquin kit foxes as well as all the other listed species in the plan (USFWS 1998). The immense challenge will be to collect the intensive demographic data necessary to complete such an analysis.

BIOPOLITICS AND ATTITUDES

In the United States, politics pervades virtually all aspects of society, and this is especially true when it comes to the issue of endangered species. Few issues generate as much passion, both pro and con, as this one, and this affects conservation and recovery efforts for San Joaquin kit foxes as it does virtually all listed species.

California has been a consistent leader in environmental protection and conservation. In his book *After the Grizzly* (Figure 4.3), Peter Alagona provides an excellent summary of the history of the conservation movement in California as well as the United States in general (Alagona 2013). Endangered species have usually fared well in California because of strong public support and favorable treatment by state politicians. San Joaquin kit foxes have benefited from these attitudes; however, a significant caveat is that most of the range of this species is located in the San Joaquin Valley, a region considered unsympathetic toward endangered species.

The reasons for this sentiment are multifold. Traditionally, the San Joaquin Valley region has been strongly conservative leaning. This may be a function of the fact that industrial-scale agriculture and oil and gas production have long been the predominant eco-

FIGURE 4.3. *After the Grizzly* by Peter Alagona. Photo by Brian Cypher.

nomic forces in the region, and individuals working in these industries tend to hold more conservative values. Also, despite the wealth of profits from agriculture and hydrocarbon extraction, only a small proportion of the population shares in this wealth, and overall the region is relatively poor (PPIC 2006, 2021). People struggling to pay bills and make ends meet are commonly less concerned with conservation of animal and plant species. Education levels also tend to be lower in the region (Cedar Lake Ventures 2021), and support for the conservation of species is commonly positively correlated with education (Alagona 2013).

A prime example of regional sentiments against endangered species was the "Taung Ming-Lin incident" in 1994. Ming-Lin was an absentee landowner who owned 291 ha (719 ac) just southwest of Bakersfield. In February 1994, despite having received official notification of the presence of endangered species on his property, he proceeded to begin disking the habitat, presumably in preparation for planting a crop. A member of the conservation public observed this action and reported it to the US Fish and Wildlife Service (USFWS) and California Department of Fish and Game (now the California Department of Fish and Wildlife—CDFW). Law enforcement agents responded and stopped the action. A subsequent search found at least one dead endangered Tipton kangaroo rat. Ming-Lin was subsequently fined and his tractor was confiscated. This sequence of events set off a heated response from conservative groups opposed to the Endangered Species Act (ESA) and who also claimed a violation of property rights. A rally was staged in Bakersfield in support of Ming-Lin and in opposition to the ESA. Ming-Lin's tractor was paraded through town, as was a giant effigy of a Tipton kangaroo rat in a rat trap. The culmination of the rally was a series of speeches by protagonists, including local organizations, officials, and politicians.

Typical attitudes in the region toward endangered species are further exemplified by the somewhat apathetic policies toward kit foxes by three colleges—California State University-Bakersfield (CSUB) and Bakersfield College (BC) located in Bakersfield, and Taft College (TC) located in Taft. San Joaquin kit foxes have been present on the campuses for as long as anyone can remember (Figure 4.4). The foxes not only occur on the campuses but also breed on all three. Thus, these colleges have the distinction of being the only campuses in the world where an endangered canid is present and breeds. Consequently, there are immense opportunities for these institutions of higher learning to take advantage of this unique, fortuitous situation and engage in education, research, outreach, conservation, and community relations involving kit foxes. However, all three institutions engage in almost none of the activities above.

On each campus, kit foxes are viewed as a curiosity at best and a nuisance at worst, because they occasionally dig dens in inconvenient locations (e.g., landscape planter boxes), including areas where the institutions have wanted to construct new buildings or other facilities. The administrations at the institutions seem to prefer that the foxes not be present on campus. However, in fairness, the institutions have always granted requests by external groups (particularly the CSUS ESRP) to conduct research and conservation activities on their campuses. The CSUB police have been especially helpful and commonly provide excellent information on current fox activities

FIGURE 4.4. San Joaquin kit foxes on the campuses of (a) California State University-Bakersfield, (b) Bakersfield College, and (c) Taft College. Photos by (a) Brian Cypher, (b) Tory Westall, and (c) Nicole Deatherage.

A

B

C

on campus. Staff at TC have assisted with research activities when possible by opening traps in the evening, reporting captures, reporting observations of foxes with mange, and generally looking out for foxes on campus. Unfortunately, the institutions have not taken advantage of the educational opportunities offered by the presence of the foxes on campus. On a positive note, in 2021, some research efforts involving undergraduate and graduate students were initiated by a new faculty member at CSUB.

Attitudes, particularly in the growing metropolitan areas, have been moderating somewhat with time. A significant motivation behind this trend interestingly lies in simple economics, particularly as it relates to retirement. An increasing number of people from more environmentally sympathetic areas, particularly the Los Angeles basin and the San Francisco Bay area, have been relocating to cities such as Fresno and Bakersfield, largely because of a much lower cost of living in these cities compared with the cities of origin for these new retirees. They have brought their attitudes along with them, and this has noticeably diluted anti-endangered species sentiments in the region. However, the region is still far from being "endangered species friendly." Interestingly, the attitudes of individuals in Bakersfield appear to be significantly influenced toward a more positive view of kit foxes just by catching glimpses of the animals (Bjurlin and Cypher 2005).

Another immense challenge for the conservation of rare species in the San Joaquin Valley region is simple competition for space. The vast majority of the land in the San Joaquin Valley, and thus the habitat upon which species depend for their existence, is privately owned. Decades ago, most of this land was converted to other uses, such as agriculture, oil and gas production, and urban development. By the time it became evident that San Joaquin kit foxes and other species were imperiled, relatively little natural habitat remained, particularly on the floor of the San Joaquin Valley. Even Grinnell noted that considerable habitat had already been converted in the early part of the twentieth century (Grinnell et al. 1937). Kelly et al. (2005) reported that only 36% of arid shrublands and 35% of arid grasslands remained in the San Joaquin Valley by the early 2000s. Consequently, the burden of endangered species conservation fell squarely on these remaining lands and their owners. The fairness can be debated, but these owners were essentially bearing the burden caused by the destruction of habitat by so many other landowners before restrictions on such destruction were imposed by the federal and California ESAs. What largely remained were grazing lands, lands owned by oil companies in anticipation that they might overlay hydrocarbon reserves, and other scattered parcels that for one reason or another had not been converted to some other use.

There were some public lands, although these made up a relatively small proportion of the whole. The US Bureau of Land Management (BLM) had considerable holdings, although many of these were located in regional mountain ranges, particularly the Coast Ranges that were suboptimal for valley floor species such as San Joaquin kit foxes. One particularly important exception was the Carrizo Plain where the BLM in collaboration with other conservation organizations conserved a large area of natural lands that became the Carrizo Plain National Monument in 2001 (see History of Conservation Efforts).

Otherwise, the vast majority of habitat conservation in the San Joaquin Valley has been driven by regulatory requirements, primarily under federal and California ESAs. In essence, proponents of development projects have been required to conserve habitat in exchange for authorization to destroy some habitat in constructing their projects. However, this is a very simplistic description of what is actually a complex situation. For example, neither the USFWS nor the CDFW or any other regulatory agency can require a project proponent to conserve habitat. That said, a precedent has been established whereby proponents "offer" to conserve habitat as part of a package of mitigation actions, and such conservation has become a successful strategy for facilitating permits necessary to proceed with development projects. Providing endowments to support long-term management and administration of the conservation lands further facilitates permitting. One other note is that an early assumption was that habitat quality on conserved lands could be enhanced to increase carrying capacity and thereby compensate for the loss of individuals (or carrying capacity) on the developed lands; however, little to no enhancement has ever been conducted on the conserved lands and this assumption is no longer a consideration in habitat conservation efforts.

The conservation mechanisms have been varied. In some cases, lands with habitat are purchased and title is transferred to a conservation organization, usually accompanied by endowments for long-term management and administration. In other cases, proponents retain ownership of their properties but place permanent conservation easements on lands with habitat. This practice is particularly common among oil companies because the HCPs typically allow the companies to retain limited development rights on the conserved lands. The end result is that significant quantities of good-quality habitat have been conserved in perpetuity.

This conservation of habitat has been conducted under the auspices of a Biological Opinion or Habitat Conservation Plan issued by the USFWS or an Incidental Take Permit issued by the CDFW. The development projects involved include hydrocarbon (e.g., oil and gas) extraction activities, urban development, water banking operations, and more recently, solar farm construction.

There is yet another significant challenge. All the above discussion on habitat conservation in the San Joaquin Valley naturally begs the question, "where are the conservation organizations?" These organizations are conspicuous by their absence. Two quotes from the chapter on San Joaquin kit foxes in *After the Grizzly* by Peter Alagona (2013) eloquently and succinctly summarize the situation:

> *The kit fox lives in a privatized, industrial landscape devoid of grandeur, bereft of beauty, lacking in charm.* (176)

> *The San Joaquin Valley has no sublime wonder; it was an industrial landscape, and at least in the view of many conservationists, it was ugly.* (180)

Some might take offense at this characterization. The reality is that the San Joaquin ecoregion generally lacks the more obvious grandeur and beauty of the surrounding mountains (e.g., Sierra Nevada, Coast Ranges) and nearby Mojave Desert, and the central coast region. Those who have spent considerable time or even careers working in the valley know that the region can have its moments (e.g., a few weeks in the spring of years with adequate precipitation in which many parts of the valley are transformed into spectacular native flower fields) and that it supports a diversity of fascinating animals and plants. However, these qualities are both ephemeral and subtle, essentially constituting an acquired appreciation. Finally, regardless of any personal sentiments, it's clear that the valley has generally been overlooked by conservation organizations known for their habitat acquisition and preservation efforts elsewhere.

The reality is that the arid scrub and grassland habitats in the San Joaquin Valley, upon which the San Joaquin kit fox and many other species depend for their existence, lack the aesthetic attributes that generate enthusiasm for conservation, particularly among groups that rely heavily on donors to fund their efforts. One very notable exception is The Wildlands Conservancy (TWC), which began acquiring land at the southern end of the San Joaquin Valley in the 1990s. Although the acquisitions were part of an effort to provide a corridor of habitat from the Mojave Desert over to the coast, one stated objective was to conserve habitat for valley floor species, including the San Joaquin kit fox. To date, approximately 40,450 ha (100,000 ac) of land, including about 8100 ha (20,000 ac) of valley floor habitat, have been acquired and are included in the TWC's Wind Wolves Preserve.

Biopolitics and the challenges described above will continue to affect conservation efforts in the San Joaquin Valley for San Joaquin kit foxes and many other species; however, considerable progress has been made despite these challenges. It is yet to be determined whether this progress is sufficient, and if not, then how much more is necessary to reduce the probability of extinction for kit foxes and other species to an acceptable level?

RESEARCH AND CONSERVATION NEEDS

Considerable research and conservation efforts have been directed toward the San Joaquin kit fox as a consequence of being a charismatic rare species (see History of Research Efforts and History of Conservation Efforts). The plethora of investigations has provided a solid foundational understanding of this animal. Aiding this understanding is the fact that kit foxes are similar in biology and ecology to a number of other species (e.g., swift fox, corsac fox, arctic fox, red fox), and gaps in data on kit foxes can be filled to some extent by inferences from data on these other species. As with any species or ecological question, having more data is always desirable, but resources for research are of course finite. In addition, the many conservation efforts, particularly the conservation of high-quality habitat, have helped lower the extinction risk for several key populations as well as the species overall. A few additional research and conservation efforts could be extremely beneficial in advancing our understanding of San Joaquin kit foxes and furthering conservation and recovery efforts.

Conserve Additional Habitat

Immense progress has been achieved in conserving habitat (see History of Conservation Efforts). As a result, the Carrizo Plain and Panoche Valley core areas are likely quite secure. Substantial acreage has also been conserved in the western Kern County core area, but protecting considerably more habitat in this area is likely necessary to truly secure the resident kit fox population. Thousands of acres of high-quality habitat remain unprotected, and the vast majority of these lands are owned by oil companies and private landowners, the latter using them primarily for grazing.

Habitat conservation is also needed in satellite population areas, with two warranting particular mention. One is the Cuyama Valley in San Luis Obispo County. This area is located just west of the Carrizo core area and is separated from this region by the Caliente Range (see Figure 2.3). Almost all the land in the Cuyama Valley is in private ownership, and much of it has been converted to agricultural uses; however, considerable acreage is used for grazing, and foxes occur on these lands. As recently as 2020, kit foxes were still observed in the Cuyama Valley and pups were even observed, indicating that reproduction was occurring. Any kit fox population is essential for long-term conservation and recovery. However, because of its relative isolation on the extreme southwestern margin of the range, this population could include unique genetic alleles not found in other populations, thereby making it even more valuable.

Another satellite population occurs in western Merced County (see Figure 2.3). In many respects, this population is analogous to the one in the Cuyama Valley, in that it is on the margin of the range (in this case, the northwest margin), it is somewhat isolated from other populations, and virtually all the suitable kit fox habitat is on private land. This area differs from the Cuyama Valley area in that most of the land is still being used for grazing and very little has been tilled. In addition, there has been interest on the part of some landowners in conserving the land. A small but persistent kit fox population occurs in this region. Seemingly, this area would be ripe for conservation efforts, particularly conservation easements that permit current land uses to continue. As with the Cuyama Valley population, this population could be valuable from a conservation genetics perspective. Recently, there has been some progress on conservation efforts with the establishment of a 971 ha (2400 ac) conservation bank by Westervelt Ecological Services, and another approximately 971 ha (2400 ac) set aside as mitigation for the Wright Solar Project.

Comprehensive Population Viability Analysis

The recovery plan for San Joaquin kit foxes and associated species (USFWS 1998) identifies several recovery tasks as well as criteria for downlisting/delisting each listed species. In general, these criteria require a certain number of populations of each species be permanently conserved in specific regions or locations. In essence, the underlying rationale is that if most of what is left is conserved, the species will persist. Unfortunately, there is no evidence to indicate that this will be sufficient to prevent the species from eventually going extinct. The authors of the plan recognized this shortcoming, and at the end of the recovery task section, tasks for virtually all the listed species call for conducting a population viability analysis, also known as a PVA. PVAs are essentially models, and their accuracy is highly dependent on the underlying assumptions and the data available to input into the model. They can be extremely useful in assessing the adequacy of conservation strategies, and in particular, the relative efficacy of alternate strategies. Therefore, these are important tools in conservation and recovery efforts.

A comprehensive PVA has not been conducted for the San Joaquin kit fox, but more specifically, a metapopulation viability analysis (MVA) needs to be conducted. San Joaquin kit foxes currently persist in a metapopulation, with some populations separated by considerable distance. Demographic data, critical for conducting an MVA, are available for the three core populations and for several of the satellite populations. Unfortunately, for some of the populations, just 1 year of data are available, meaning that natural variation in demographic parameters cannot be factored into the calculations. However, even if imperfect, an MVA will be extremely informative regarding the prospects for persistence of the species under current conservation strategies and whether more extreme conservation measures might be necessary. Such an analysis would also provide estimates of extinction probability for individual populations, thereby identifying populations where conservation efforts need to be enhanced.

Outreach

San Joaquin Valley species in general have suffered from a lack of outreach efforts, for numerous reasons. These reasons can be broadly categorized as (1) lack of resources, (2) lack of significant environmental advocacy, and (3) low public receptivity. The lack of resources includes absence of agency presence, as well as overall lack of personnel and funding. The nearest USFWS office is located in Sacramento (well outside the range of the San Joaquin kit fox), save for a lone staff-level position based in Fresno. The nearest CDFW office is in Fresno, and several staff are also based in Bakersfield. However, both agencies have had limited resources for any sort of large-scale, far-reaching outreach efforts. In addition, relatively little advocacy by nongovernmental groups occurs in the San Joaquin Valley. This is due partly to lack of any major groups based in the valley. In addition, although charismatic in their own right, none of the species is sufficiently charismatic to attract widespread public interest (as do California condors, sea otters, giant pandas, and gray wolves). Additionally, as discussed, the natural habitats that support kit foxes and co-occurring species are arguably unappealing and unattractive, save for a few short weeks in wetter years when flower displays can be outstanding. This lack of charismatic and aesthetic appeal results in relatively low interest in the species and their ecosystems and also diminishes their value for fundraising, a prime motivation for many conservation-oriented nongovernmental organizations (NGOs). Finally, the residents of the valley are not the most receptive to information or appeals regarding endangered species. As mentioned (see Biopolitics and Attitudes), the valley is relatively conservative in its political leanings, and relatively low education and income levels exacerbate the situation. Consequently, other issues tend to take precedence in the lives of residents, and endangered species concerns are typically not a priority.

Despite these immense challenges, enhanced outreach efforts could only help kit foxes and other species. Efforts that target younger age classes (e.g., elementary schoolchildren) could help sow seeds of conservation interest and ethics that might sprout and take root in at least a portion of children. In addition, as described earlier, urban kit foxes are fantastic ambassadors for their species in that people who have observed foxes are more likely to support conservation of foxes in urban and also non-urban environments. Thus, outreach targeting city residents might bolster support for kit fox conservation throughout the range.

The California Living Museum (CALM), a small zoo located on the outskirts of Bakersfield, has been doing its best to address the need for outreach. Since opening in 1983, CALM has almost continuously displayed San Joaquin kit foxes. Most have been animals found injured or orphaned or illegally kept as pets, and then determined to be non-releasable. On occasion, CALM has had human-habituated kit foxes that were used as "ambassador animals"; they were taken to outreach events, used in presentations at the zoo, and other activities (Figure 4.5). In addition to displaying animals, CALM also conducts considerable outreach and education, engages in conservation activities, and treats and rehabilitates sick and injured wildlife. CALM has played a significant and crucial role in the battle against sarcoptic mange in the Bakersfield kit fox population. Outreach efforts in addition to those conducted by CALM could be immensely helpful.

A final note is that in the absence of coordinated agency efforts or the presence of major environmental NGOs, significant outreach may best be achieved by a dedicated organization, such as a "Friends of the San Joaquin Kit Fox" group. Such a group and its efforts could easily be modeled after a number of other successful groups, including Friends of the Island Fox, Friends of the Sea Otter, and others. Such groups can have low operational costs with little or no bureau-

FIGURE 4.5. Dash, an ambassador San Joaquin kit fox, at the California Living Museum in Bakersfield. Photo by Kourtney Hall.

cracy, and they also tend to be locally based and therefore better positioned to secure local public support.

Other research and conservation needs likely exist, but these are some of the more substantial ones that could have significant impacts on the conservation and recovery of San Joaquin kit foxes. Updating the recovery plan that includes San Joaquin kit foxes (USFWS 1998) might help to identify additional needs and formalize steps to implement them.

Conclusion

THE FUTURE

The future for San Joaquin kit foxes is still uncertain. A number of different factors, even operating singularly, could determine the ultimate fate of this taxon. Most have been discussed in detail previously and are summarized here.

Habitat loss and degradation has been the primary factor responsible for the decline of the San Joaquin kit fox. As of this writing, this loss and degradation continues at a pace that could lead to the extinction of co-occurring rare species (e.g., Tipton kangaroo rats; Cypher, Phillips, et al. 2021) in the near term, and to the erosion and loss of the San Joaquin kit fox populations to a point at which the overall metapopulation is no longer viable in the long term. Then it would just be a matter of time before stochastic processes or catastrophic events lead to the eventual extirpation of each of the remaining populations, resulting in extinction for this taxon. It certainly is true that an increasing proportion of the remaining kit fox habitat has been conserved through acquisition, conservation easements, and other processes, but whether this will be enough to preserve the species in the long term is uncertain.

Plausible scenarios include the continued loss of habitat, but also potentially include the opposite: an increase in the amount of available habitat. The continued rapid growth of the human population within the range of the San Joaquin kit fox could increase the demand for land for housing, commerce, industry, recreation, and other uses. For example, from 2000 to 2017, the population of the city of Bakersfield increased by more than 55%, and the urbanized area increased by 32% (Visit Bakersfield 2021). The many other urban areas in the San Joaquin Valley are also growing in population and land area at similar rates. Habitat could also continue to be converted to agricultural uses, although this industry is strongly driven by demand, economics, and water availability. The probability that water supplies will increase is likely low, particularly as the Sierra snowpack continues to decline as a result of climate change. However, water use per acre is decreasing as farmers move to new crops requiring less water, research develops varieties of current crops requiring less water, and technological advances (e.g., better drip irrigation) make it possible to grow the same crops with less water. All these advances could de facto result in an increase in water availability and concomitant additional conversion of habitat to crop land. Finally, other industries or practices could come along that result in conversion of more habitat. For example, 20 years ago, the explosive growth in solar energy production that has occurred within the range of the San Joaquin kit fox was on virtually no one's radar as a potential threat to the species. Fortunately, in the case of solar energy plants, it appears that they can be made "fox friendly"; however, that may not be true of the next "big thing" that comes along. For example, in the late decades of the twentieth century, the San Joaquin Valley suddenly became a mecca for mega-dairies, requiring many acres to hold thousands of cows at each facility, and even many times more acres to grow food crops (e.g., alfalfa, silage) to feed the cows. As another example, massive warehouses and distribution centers have recently been proliferating in the San Joaquin Valley, each requiring tens of acres for the buildings and associated infrastructure. So demands for land will likely continue within the range of the San Joaquin kit fox.

Other potential scenarios could result in some currently converted lands reverting back to usable habitat for kit foxes. As frequently mentioned here, water availability in the San Joaquin Valley is already becoming significantly restricted. The agricultural industry in particular is bearing the brunt of this impact. Already, farmers have been taking some lands, particularly those with less productive soils, out of production due to lack of available water. As mentioned on several occasions, the Sustainable Groundwater Management Act of 2014 will further decrease water availability as restrictions increase on the pumping of ground water. This scenario will likely result in extensive acreage eventually being taken out of agricultural production. One possibility is that these lands will be allowed to revert to natural habitat, either through natural processes or through active restoration. However, it is also possible that alternate economic or other uses will be found for some or all of these lands, and thus they will not be available for habitat. Current regulations will likely influence these events as well. For example, if landowners fear that colonization of their lands by endangered species might place immediate burdens on them or limit future uses of the land, they may take steps (e.g., periodic disking) to inhibit or prevent such colonization. Innovative strategies will be required to facilitate the reversion of lands to usable habitat.

Of further concern are unforeseen and unpredictable factors, like disease or some large-scale catastrophic event. Before about 2010, disease did not seem to be an issue for kit foxes. Then, there was the distemper event among desert kit foxes at a solar facility near Blythe, the sarcoptic mange epidemic in Bakersfield, and the occurrence of distemper at the Panoche solar site. Whether these events were just completely coincidental or a function of anthropogenic or other influences is unknown; however, if anthropogenically influenced, such influences will most certainly persist and likely increase as the human population within the range of the San Joaquin kit fox increases. Even if the events were not anthropogenically influenced but instead were the result of ecological factors or random chance, the number and size of the remaining kit fox populations has shrunk and, as a consequence, each population and the entire metapopulation are more vulnerable. Of course, climate change is lurking like the 800-pound gorilla in the back of the room, and only time will tell what impact it will have.

Numerous conservation efforts for San Joaquin kit foxes are in progress as of this writing. These include habitat conservation efforts, research efforts, and outreach efforts. Again, it is unclear whether these efforts will be sufficient to prevent San Joaquin kit foxes from going extinct. With all the current uncertainty, more aggressive and progressive approaches may be warranted, including some considered "outside the box." An example of one such idea is the introduction of kit foxes into other urban areas in the San Joaquin Valley (as

FIGURE C.1. Litter of San Joaquin kit fox pups. Photo by Christine Van Horn Job.

proposed in Kit Foxes and Urban Areas). Ultimately, the more populations there are, the lower the extinction risk for the San Joaquin kit fox metapopulation. Ideally, San Joaquin kit foxes will manage to persist in natural habitats where they are subject to natural ecological and evolutionary processes. Seeing a den of kit fox pups is exciting regardless of location, but seeing one in a natural setting (Figure C.1) is truly wonderful, as that's "the way it should be."

Even with the development and implementation of more progressive, creative conservation strategies, it is challenging to envision a scenario in which the San Joaquin kit fox metapopulation will increase to a sufficient size and state of stability such that this taxon can be removed from the Endangered Species list. Instead, it is more likely that San Joaquin kit foxes will require continued protections and conservation efforts in order to persist. Scott et al. (2005) referred to species needing such measures to avoid extinction as being "conservation reliant." In the absence of some unanticipated change in events, this will likely be the fate of the San Joaquin kit fox.

With their adaptability and mobility, kit foxes stand a better chance of hanging on compared with some other species. The loss of this San Joaquin Valley native would not only be a sad and depressing outcome, but it would also indicate poor prospects for other less adaptable and less mobile co-occurring species, and for arid ecosystems in the San Joaquin Valley in general. However, I am optimistically hopeful, hopefully not naively so, that people will continue to catch glimpses of this charismatic species for generations to come.

Appendix

COMMON AND SCIENTIFIC NAMES OF SPECIES

MAMMALS

Arctic fox (*Vulpes lagopus*)

Badger (*Taxidea taxus*)

Black bear (*Ursus americanus*)

Black rat (*Rattus rattus*)

Bobcat (*Lynx rufus*)

California ground squirrel (*Otospermophilus beechyi*)

Civet (family Viverridae)

Corsac fox (*Vulpes corsac*)

Coyote (*Canis latrans*)

Deer (*Odocoileus* spp.)

Deer mice (*Peromyscus maniculatus*)

Desert kangaroo rat (*Dipodomys deserti*)

Desert kit fox (*Vulpes macrotis arsipus*)

Domestic dog (*Canis lupus familiaris*)

Fresno kangaroo rat (*Dipodomys nitratoides exilis*)

Giant panda (*Ailuropoda melanoleuca*)

Gopher (*Thomomys bottae*)

Gray fox (*Urocyon cinereoargenteus*)

Gray wolf (*Canis lupus*)

Giant kangaroo rat (*Dipodomys ingens*)

Heermann's kangaroo rat (*Dipodomys heermanni*)

House cat (*Felis catus*)

House mouse (*Mus musculus*)

Island fox (*Urocyon littoralis*)

Kangaroo rat (*Dipodomys* spp.)

Merriam's kangaroo rat (*Dipodomys merriami*)

Mongoose (family Herpestidae)

Mountain lion (*Puma concolor*)

Norway rat (*Rattus norvegicus*)

Opossum (*Didelphis virginiana*)

Pocket mice (*Perognathus* spp. and *Chaetodipus* spp.)

Polar bear (*Ursus maritimus*)

Raccoon (*Procyon lotor*)

Red fox (*Vulpes vulpes*)

San Joaquin antelope squirrel (*Ammospermophilus nelsoni*)

San Joaquin kangaroo rat (*Dipodomys nitratoides*)

San Joaquin kit fox (*Vulpes macrotis mutica*)

Sea otter (*Enhydra lutris*)

Short-nosed kangaroo rat (*Dipodomys nitratoides brevinasus*)

Striped skunk (*Mephitis mephitis*)

Swift fox (*Vulpes velox*)

Tipton kangaroo rat (*Dipodomys nitratoides nitratoides*)

White-tailed antelope squirrel (*Ammospermophilus leucurus*)

BIRDS

Burrowing owl (*Athene cunicularia*)

California condor (*Gymnogyps californianus*)

Canada goose (*Branta canadensis*)

Golden eagle (*Aquila chrysaetos*)

Great-horned owl (*Bubo virginianus*)

Le Conte's thrasher (*Toxostoma lecontei*)

REPTILES AND AMPHIBIANS

Blunt-nosed leopard lizard (*Gambelia sila*)

Desert spiny lizard (*Sceloporus magister*)

Desert tortoises (*Gopherus agassizii*)

Glossy snake (*Arizona elegans*)

Long-nosed leopard lizard (*Gambelia wislizenii*)

Pacific rattlesnake (*Crotalus oreganus*)

Side-blotched lizard (*Uta stansburiana*)

Western toad (*Anaxyrus boreas*)

INVERTEBRATES

Beetles (order Coleoptera)
Cockroaches (order Blattodea)
Crickets (order Orthoptera)
Earwigs (order Dermaptera)
Grasshoppers (order Orthoptera)
Jerusalem crickets (family Stenopelmatidae)
Moths (order Lepidoptera)
Scorpions (order Scorpiones)

PLANTS

Alkali goldenbush (*Isocoma acradenia*)
Allscale saltbush (*Atriplex polycarpa*)
Arabian grass (*Schismus arabicus*)
Bakersfield cactus (*Opuntia basilaris* var. *treleasei*)
Beavertail prickly-pear (*Opuntia basilaris*)
Bromes (*Bromus* spp.)
California jewelflower (*Caulanthus californicus*)
Creosote bush (*Larrea tridentata*)
Desert saltbush (*Atriplex polycarpa*)
Fiddlenecks (*Amsinckia* spp.)
Fivehook bassia (*Bassia hyssopifolia*)
Giant sequoia (*Sequoiadendron giganteum*)
Goldfields (*Lasthenia* spp.)
Hoover's woolly-star (*Eriastrum hooveri*)
Jackass clover (*Wislizenia refracta*)
Jointfir (*Ephedra californica*)
Kern mallow (*Eremalche parryi* ssp. *kernensis*)
Mustards (family Brassicaceae)
Oak (*Quercus* spp.)
Phacelias (*Phacelia* spp.)
Red brome (*Bromus madritensis* ssp. *rubens*)
Red-stemmed filaree (*Erodium cicutarium*)
San Joaquin woolly-threads (*Monolopia* (=*Lembertia*) *congdonii*)
Spinescale saltbush (*Atriplex spinifera*)
Spiny saltbush (*Atriplex spinifera*)
Tarweeds (family Asteraceae)
Tumbleweeds (*Salsola* spp.)
Wild barley (*Hordeum* spp.)
Wild oats (*Avena* spp.)

LITERATURE CITED

Alagona, P. S. 2013. After the grizzly: Endangered species and the politics of place in California. Berkeley: University of California Press.

Alderton, D. 1994. Foxes, wolves and wild dogs of the world. New York: Facts on File, Inc.

Althouse and Meade, Inc. 2010. Topaz Solar Farms San Joaquin kit fox mitigation and monitoring plan. Paso Robles, CA: Althouse and Meade, Inc.

Anderson, E. M., and M. J. Lovallo. 2003. Bobcat and lynx. In G. A. Feldhamer, B. C. Thompson, and J. A. Chapman, eds., Wild mammals of North America: Biology, management, and conservation, 2nd ed., pp. 758–786. Baltimore, MD: Johns Hopkins University Press.

Anderson, R. L., L. K. Spiegel, and K. M. Kakiba-Russell. 1991. Southern San Joaquin Valley ecosystems protection program: Natural lands inventory and maps. Sacramento: California Energy Commission.

Ausband, D. E., and E. A. Ausband. 2006. Observations of interactions between swift fox and American badger. Prairie Naturalist 38:63–64.

Baker, H. G. 1978. Invasion and replacement in California and Neotropical grasslands. In J. R. Wilson, ed., Plant relations in pastures, pp. 368–384. Melbourne, Australia: Commonwealth Scientific and Industrial Research Organization.

Balestreri, A. N. 1981. Status of the San Joaquin kit fox at Camp Roberts, California, 1981. Fort Ord, CA: Department of the Army.

Bean, W. T., L. Charles, and L. Larios. 2020. Carrizo Plain ecosystem project annual report (2018 & 2019). San Luis Obispo: California Polytechnic State University.

Beedy, E. C., V. K. Getz, and D. A. Airola. 1987. Status of the San Joaquin kit fox (*Vulpes macrotis mutica*) in the urban Kern River Parkway, Bakersfield, California. In D. F. Williams, S. Byrne, and T. A. Rado, eds., Endangered and sensitive species of the San Joaquin Valley, California: Their biology, management, and conservation, pp. 47–53. Sacramento: California Energy Commission.

Bekoff, M. 2001. Human-carnivore interactions: adopting proactive strategies for complex problems. In J. L. Gittleman, S. M. Funk, D. Macdonald, and R. K. Wayne, eds., Carnivore conservation, pp. 179–195. Cambridge: Cambridge University Press.

Bergstrom, B. J. 2017. Carnivore conservation: Shifting the paradigm from control to coexistence. Journal of Mammalogy 98:1–6.

Berry, W. H., T. P. O'Farrell, T. T. Kato, and P. M. McCue. 1987. Characteristics of dens used by radiocollared San Joaquin kit fox (*Vulpes macrotis mutica*), Naval Petroleum Reserve #1, Kern County California. US Department of Energy Topical Report EGG 10282-2177. Springfield, VA: National Technical Information Service.

Berry, W. H., and W. G. Standley. 1992. Population trends of San Joaquin kit fox at Camp Roberts Army National Guard Training Site, California. US Department of Energy Topical Report EGG 10617-2155. Springfield, VA: National Technical Information Service.

Best, T. L. 1991. *Dipodomys nitratoides*. Mammalian Species 381:1–7.

Bjurlin, C. D., and B. L. Cypher. 2003. Effects of roads on San Joaquin kit foxes: A review and synthesis of existing data. *In* C. L. Irwin, P. Garrett, and K. P. McDermott, eds., Proceedings of the International Conference on Ecology and Transportation, pp. 397–406. Raleigh: Center for Transportation and the Environment, North Carolina State University.

Bjurlin, C. D., and B. L. Cypher. 2005. Encounter frequency with the urbanized San Joaquin kit fox correlates with public beliefs and attitudes toward the species. Endangered Species Update 22:107–115.

Bjurlin, C. D., B. L. Cypher, C. M. Wingert, and C. L. Van Horn Job. 2005. Urban roads and the endangered San Joaquin kit fox. Fresno: California State University-Stanislaus, Endangered Species Recovery Program.

Bradley, C. A., and S. Altizer. 2007. Urbanization and the ecology of wildlife diseases. Trends in Ecology and Evolution 22:95–102.

Brattstrom, B. H. 1953. Records of Pleistocene reptiles from California. Copeia 1953:174–179.

Bremner-Harrison, S., and B. L. Cypher. 2007. Feasibility and strategies for translocating San Joaquin kit foxes to vacant or restored habitats. Fresno: California State University-Stanislaus, Endangered Species Recovery Program.

Bremner-Harrison, S., and B. L. Cypher. 2011. Reintroducing San Joaquin kit foxes to vacant or restored lands: Identifying optimal source populations and candidate foxes. Fresno: California State University-Stanislaus, Endangered Species Recovery Program.

Bremner-Harrison, S., B. L. Cypher, C. M. Fiehler, A. P. Clevenger, and D. Hacker. 2007. Use of highway crossing structures by kit foxes. Fresno: California State University-Stanislaus, Endangered Species Recovery Program.

Bremner-Harrison, S., B. L. Cypher, and S. W. R. Harrison. 2013. An investigation into the effect of individual personality on re-introduction success, examples from three North American fox species: Swift fox, California Channel Island fox and San Joaquin kit fox. *In* P. S. Soorae, ed., Global re-introduction perspectives: 2013, further case-studies from around the globe, pp. 152–158. Gland, Switzerland: IUCN/SSC Re-introduction Specialist Group and Abu Dhabi Environment Agency.

Briden, L. E., M. Archon, and D. L. Chesemore. 1992. Ecology of the San Joaquin kit fox (*Vulpes macrotis mutica*) in Western Merced County, California. *In* D. F. Williams, S. Byrne, and T. A. Rado, eds., Endangered and sensitive species of the San Joaquin Valley, California: Their biology, management, and conservation, pp. 81–87. Sacramento: California Energy Commission.

Brown, J. H., and B. A. Harney. 1993. Population and community ecology of heteromyid rodents in temperate habitats. *In* H. H. Genoways and J. H. Brown, eds., Biology of the Heteromyidae, American Society of Mammalogists, Special Publication No. 10, pp. 618–651. Provo, UT: Brigham Young University.

Burt, W. H. 1943. Territoriality and home range concepts as applied to mammals. Journal of Mammalogy 24:346–352.

Butterfield, H. S., D. Cameron, E. Brand, M. Webb, E. Forsburg, M. Kramer, E. O'Donoghue, and L. Crane. 2013. Western San Joaquin Valley least conflict solar assessment. San Francisco: The Nature Conservancy.

Butterfield, H. S., T. R. Kelsey, and A. K. Hart. 2021. Rewilding agricultural landscapes: A California study in rebalancing the needs of people and nature. Washington, DC: Island Press.

Cameron, D. R., B. S. Cohen, and S. A. Morrison. 2012. An approach to enhance the conservation-compatibility of solar energy development. PLoS ONE 7(6): e38437.

Careau, V., J.-F. Giroux, G. Gauthier, and D. Berteaux. 2008. Surviving on cached foods—the energetics of egg-caching by arctic foxes. Canadian Journal of Zoology 86:1217–1223.

Case, T. J., and M. E. Gilpin. 1974. Interference competition and niche theory. Proceedings of the National Academy of Science 71:3073–3077.

California Energy Commission. 2020. California solar energy statistics and data. Accessed January 12, 2023. https://ww2.energy.ca.gov/almanac/renewables_data /solar/index_cms.php

Cedar Lake Ventures. 2021. Educational Attainment in California. Data from US Census Bureau. Accessed January 12, 2023. https://statisticalatlas.com/state /California/Educational-Attainment#data-map/county

Clark, H. O., Jr. 2007. Marking of novel objects by kit foxes. California Fish and Game 93:103–106.

Clark, H. O., Jr., D. P. Newman, and S. I. Hagen. 2007. Analysis of San Joaquin kit fox element data with the California Diversity Database: A case for data reliability. Transactions of the Western Section of The Wildlife Society 43:37–42.

Clark, H. O., Jr., R. M. Powers, K. L. Uschyk, and R. K. Burton. 2015. Observations of antagonistic and nonantagonistic interactions between the San Joaquin kit fox (*Vulpes macrotis mutica*) and the American badger (*Taxidea taxus*). Southwestern Naturalist 60:106–110.

Clark, H. O., Jr., G. D. Warrick, B. L. Cypher, P. A. Kelly, D. F. Williams, and D. E. Grubbs. 2005. Competitive interactions between endangered kit foxes and nonnative red foxes. Western North American Naturalist 65:153–163.

Clevenger, A. P., Kociolek, A. V., and B. L. Cypher. 2010. Effects of four-lane highways on desert kit fox and swift fox: Inferences for the San Joaquin kit fox population. Bozeman: Montana Western Transportation Institute, Montana State University.

Conover, A. 2001. The little foxes. Smithsonian 32(5): 42–50.

Constable, J. L., B. L. Cypher, S. E. Phillips, and P. A. Kelly. 2009. Conservation of San Joaquin kit foxes in western Merced County, California. Fresno: California State University-Stanislaus, Endangered Species Recovery Program.

Coonan, T. J., C. A. Schwemm, and D. K. Garcelon. 2010. Decline and recovery of the Island fox: A case study for population recovery. New York: Cambridge University Press.

Crandall, K. A., O. R. P. Bininda-Emonds, G. M. Mace, and R. K Wayne. 2000. Considering evolutionary processes in conservation biology. Trends in Ecology and Evolution 15:290–295.

Curtis, P. D., and J. Hadidian. 2010. *In* S. D. Gehrt, S. P. D. Riley, and B. L. Cypher, eds., Urban carnivores: Ecology, conflict, and conservation, pp. 201–211. Baltimore, MD: Johns Hopkins University Press.

Cypher, B. L. 1993. Food item use by three sympatric canids in southern Illinois. Transactions of the Illinois State Academy of Science 86:139–144.

Cypher, B. L. 2000. Effects of roads on San Joaquin kit foxes: A review and synthesis of literature. Fresno: California State University-Stanislaus, Endangered Species Recovery Program.

Cypher, B. L. 2001. Spatiotemporal variation in rodent abundance in the San Joaquin Valley, California. Southwestern Naturalist 46:66–75.

Cypher, B. L. 2003. Foxes. *In* G. A. Feldhamer, B. C. Thompson, and J. A. Chapman, eds., Wild mammals of North America: Biology, management, and conservation, 2nd ed., pp. 511–546. Baltimore, MD: The Johns Hopkins University Press.

Cypher, B. L. 2010. Kit foxes. *In* S. D. Gehrt, S. P. D. Riley, and B. L. Cypher, eds., Urban carnivores: Ecology, conflict, and conservation, pp. 49–60. Baltimore, MD: The Johns Hopkins University Press.

Cypher, B. L., and A. D. Brown. 2006. Demography and ecology of endangered San Joaquin kit foxes at the Bena Landfill, Kern County, California. Fresno: California State University-Stanislaus, Endangered Species Recovery Program.

Cypher, B. L., and N. Frost. 1999. Condition of San Joaquin kit foxes in urban and exurban habitats. Journal of Wildlife Management 63:930–938.

Cypher, B. L., and J. H. Scrivner. 1992. Coyote control to protect endangered San Joaquin kit foxes at the Naval Petroleum Reserves, California. Proceedings of the Vertebrate Pest Conference 15:42–47.

Cypher, B. L., and K. A. Spencer. 1998. Competitive interactions between coyotes and San Joaquin kit fox. Journal of Mammalogy 79:204–214.

Cypher, B. L., C. D. Bjurlin, and J. L. Nelson. 2009. Effects of roads on endangered San Joaquin kit foxes. Journal of Wildlife Management 73:885–893.

Cypher, B. L., C. D. Bjurlin, and J. L. Nelson. 2018. Comparison of rabbit abundance survey techniques in arid habitats. California Fish and Game 104:19–27.

Cypher, B. L., B. B. Boroski, R. K. Burton, D. E. Meade, S. E. Phillips, P. Leitner, E. C. Kelly, T. L. Westall, and J. Dart. 2021. Photovoltaic solar farms in California: Can we have renewable electricity and our species, too? California Fish and Wildlife Journal 107:231–248.

Cypher, B. L., H. O. Clark, Jr., P. A. Kelly, C. Van Horn Job, G. W. Warrick, and D. F. Williams. 2001. Interspecific interactions among mammalian predators: Implications for the conservation of endangered San Joaquin kit foxes. Endangered Species UPDATE 18:171–174.

Cypher, B. L., E. A. Cypher, D. J. Germano, E. N. Tennant, and L. R. Saslaw. 2021. Rewilding through reintroduction. *In* H. S. Butterfield, T. R. Kelsey, and A. K. Hart, eds. Rewilding agricultural landscapes: A California study in rebalancing the needs of people and nature, pp. 95–108. Washington, DC: Island Press.

Cypher, B. L., N. A. Deatherage, T. L. Westall, and E. C. Kelly. 2022. Intraguild competition between endangered kit foxes and a novel predator in a novel environment. Animals 12:2727.

Cypher, B. L., N. A. Deatherage, T. L. Westall, E. C. Kelly, and S. E. Phillips. 2023. Potential habitat and carrying capacity of endangered San Joaquin kit foxes in an urban environment: Implications for conservation and recovery. Urban Ecosystems 26(1): 173–183.

Cypher, B. L., C. M. Fiehler, T. L. Westall, C. L. Van Horn Job, and E. C. Kelly. 2014. San Joaquin kit fox conservation in the northern Carrizo Plain: Baseline demographic and ecological attributes. Turlock: California State University–Stanislaus, Endangered Species Recovery Program.

Cypher, B. L., M. E. Koopman, and D. R. McCullough. 2001. Space use and movements by kit fox family members. Transactions of the Western Section of the Wildlife Society 37:84–87.

Cypher, B. L., J. D. Murdoch, and A. D. Brown. 2021. Artificial dens for the conservation of San Joaquin kit foxes. California Fish and Wildlife Journal Special CESA Issue:416–437.

Cypher, B. L., S. C. McMillin, T. L. Westall, C. Van Horn Job, R. C. Hosea, B. J. Finlayson, and E. C. Kelly. 2014. Rodenticide exposure among endangered kit foxes relative to habitat use in an urban landscape. Cities and the Environment 7(1): Article 8.

Cypher, B. L., S. E. Phillips, and P. A. Kelly. 2013. Quantity and distribution of suitable habitat for endangered San Joaquin kit foxes: Conservation implications. Canid Biology and Conservation 16:25–31.

Cypher, B. L., S. E. Phillips, T. L. Westall, E. N. Tennant, L. R. Saslaw, E. C. Kelly, and C. L. Van Horn Job. 2021. Conservation of endangered Tipton kangaroo rats (*Dipodomys nitratoides nitratoides*): Status surveys, habitat suitability, and conservation recommendations. California Fish and Wildlife Journal Special CESA Issue:382–397.

Cypher, B. L., J. L. Rudd, T. L. Westall, L. W. Woods, N. Stephenson, J. E. Foley, D. Richardson, and D. L. Clifford. 2017. Sarcoptic mange in endangered kit foxes: Case histories, diagnoses, and implications for conservation. Journal of Wildlife Diseases 53:46–53.

Cypher, B. L., K. A. Spencer, T. L. Westall, and D. E. Meade. 2019. Golden eagle predation on endangered San Joaquin kit foxes. Western North American Naturalist 79:556–563.

Cypher, B. L., and C. L. Van Horn Job. 2012. Management and conservation of San Joaquin kit foxes in urban environments. Proceedings of the Vertebrate Pest Conference 25:347–352.

Cypher, B. L., and G. D. Warrick. 1993. Use of human-derived food items by urban kit foxes. Transactions of the Western Section of The Wildlife Society 29:34–37.

Cypher, B. L., G. D. Warrick, M. R. M. Otten, T. P. O'Farrell, W. H. Berry, C. E. Harris, T. T. Kato, P. M. McCue, J. H. Scrivner, and B. W. Zoellick. 2000. Population dynamics of San Joaquin kit foxes at the Naval Petroleum Reserves in California. Wildlife Monographs 145:1–43.

Cypher, B. L., Westall, T. L., E. C. Kelly, and N. A. Deatherage. 2023. Demographic and ecological responses of endangered San Joaquin kit foxes to the Panoche Valley Solar Farm. Turlock: California State University-Stanislaus, Endangered Species Recovery Program.

Cypher, B. L., T. L. Westall, E. C. Kelly, N. A. Deatherage, C. L. Van Horn Job, and

L. R. Saslaw. 2022. Demographic and ecological patterns of endangered San Joaquin kit foxes in the Carrizo Plain National Monument. Turlock: California State University-Stanislaus, Endangered Species Recovery Program.

Cypher, B. L., T. L. Westall, E. C. Kelly, N. A. Deatherage, G. D. Warrick, and L. Spiegel. 2021. Fostering orphaned pups of endangered San Joaquin kit foxes (*Vulpes macrotis mutica*): Three case studies. Journal of Wildlife Rehabilitation 41(3): 23–29.

Cypher, B. L., T. L. Westall, K. A. Spencer, D. E. Meade, E. C. Kelly, J. Dart, and C. L. Van Horn Job. 2019. Response of San Joaquin kit foxes to Topaz Solar Farms: Implications for conservation of kit foxes. Turlock: California State University-Stanislaus, Endangered Species Recovery Program.

Cypher, B. L., T. L. Westall, C. L. Van Horn Job, and E. C. Kelly. 2014. San Joaquin kit fox conservation in a satellite habitat area. Turlock: California State University-Stanislaus, Endangered Species Recovery Program.

Deatherage, N. A., B. L. Cypher, J. Murdoch, T. L. Westall, E. C. Kelly, and D. J. Germano. 2021. Urban landscape attributes affect occupancy patterns of the San Joaquin kit fox during an epizootic. Pacific Conservation Biology 27:256–266.

Deatherage, N. A., E. C. Kelly, T. L. Westall, S. Adams, J. L. Rudd, D. L. Clifford, and B. L. Cypher. 2023. Case study: Cross-fostering of an endangered San Joaquin kit fox pup (*Vulpes macrotis mutica*). Journal of Wildlife Rehabilitation: in press.

Dragoo, J. W., J. R. Choate, T. L. Yates, and T. P. O'Farrell. 1990. Evolutionary and taxonomic relationships among North American arid-land foxes. Journal of Mammalogy 71:318–332.

Dragoo, J. W., and R. K. Wayne. 2003. Systematics and population genetics. *In* M. A. Sovada and L. Carbyn, eds., The swift fox: Ecology and conservation of swift foxes in a changing world, pp. 207–222. Regina, Saskatchewan: Canadian Plains Research Center, University of Regina.

Egoscue, H. J. 1956. Preliminary studies of the kit fox in Utah. Journal of Mammalogy 37:351–357.

Egoscue, H. J. 1962. Ecology and life history of the kit fox in Tooele County, Utah. Ecology 43:481–497.

Egoscue, H. J. 1975. Population dynamics of the kit fox in western Utah. Bulletin of the Southern California Academy of Science 74:122–127.

Egoscue, H. J. 1979. *Vulpes velox*. Mammalian Species 122:1–7.

Egoscue, H. J. 1985. Kit fox flea relationships on the Naval Petroleum Reserves, Kern County, California. Bulletin of the Southern California Academy of Sciences 84:127–132.

Entrix, Inc. 2010. Potential habitat of selected federally listed species: An assessment in partial fulfillment of the analysis of effect of climate change on future habitat distribution. Ventura, CA: Entrix, Inc.

Ewer, R. F. 1973. The carnivores. Ithaca, NY: Cornell University Press.

Feinstein, L. C., S. Phillips, J. Banbury, A. Hamdoun, S. C. T. Nicklisch, and B. L. Cypher. 2015. Potential impacts of well stimulation on wildlife and vegetation. *In* J. C. S. Long, J. T. Birkholzer, and L. C. Feinstein, eds., An independent scientific assessment of well stimulation in California; Volume II Potential environmental impacts of hydraulic fracturing and acid stimulations, pp. 310–373. Sacramento: California Council on Science and Technology.

Fiehler, C. M., B. L. Cypher, and L. R. Saslaw. 2017. Effects of oil and gas development on vertebrate community composition in the southern San Joaquin Valley, California. Global Ecology and Conservation 9:131–141.

Forman, R. T. T., and L. E. Alexander. 1998. Roads and their major ecological effects. Annual Review of Ecology and Systematics 29:207–231.

Frank, D., and R. Inouye. 1994. Temporal variation in actual evapotranspiration of terrestrial ecosystems: Patterns and ecological implications. Journal of Biogeography 21:401–411.

Frankham, R., J. D. Ballou, K. Ralls, M. D. B. Eldridge, M. R. Dudash, C. B. Fenster, R. C. Lacy, and P. Sunnucks. 2017. Genetic management of fragmented animal and plant populations. Oxford: Oxford University Press.

Frost, N. 2005. San Joaquin kit fox home range, habitat use, and movements in urban Bakersfield. Thesis, Humboldt State University, Arcata, CA.

Fuller, T. K., and P. R. Sievert. 2001. Carnivore demography and the consequences of changes in prey availability. In J. L. Gittleman, S. M. Funk, D. Macdonald, and R. K. Wayne, eds., Carnivore conservation, pp. 163–178. Cambridge: Cambridge University Press.

Germano, D. J., R. B. Rathbun, and L. R. Saslaw. 2001. Managing exotic grasses and conserving declining species. Wildlife Society Bulletin 29:551–559.

Germano, D. J., G. B. Rathbun, and L. R. Saslaw. 2012. Effects of grazing and invasive grasses on desert vertebrates in California. Journal of Wildlife Management 76:670–682.

Germano, D. J., G. B. Rathbun, L. R. Saslaw, B. L. Cypher, E. A. Cypher, and L. M. Vredenburgh. 2011. The San Joaquin desert of California: Ecologically misunderstood and overlooked. Natural Areas Journal 31:138–147.

Germano, D. J., and L. R. Saslaw. 2017. Rodent community dynamics as mediated by environment and competition in the San Joaquin Desert. Journal of Mammalogy 98:1615–1626.

Ginsberg, J. R., G. M. Mace, and S. Albon. 1995. Local extinction in a small and declining population: Wild dogs in the Serengeti. Proceedings of the Royal Society of London B. 262:221–228.

Gittleman, J. L. S. M. Funk, D. W. Macdonald, and R. K. Wayne. 2001. Why 'carnivore conservation'? In J. L. Gittleman, S. M. Funk, D Macdonald, and R. K. Wayne, eds., Carnivore conservation, pp. 1–8. Cambridge: Cambridge University Press.

Gittleman, J. L., and S. D. Thompson. 1988. Energy allocation in mammalian reproduction. American Zoologist 28:863–875.

Golightly, Jr., R. T., and R. D. Ohmart. 1983. Metabolism and body temperature of two desert canids: Coyotes and kit foxes. Journal of Mammalogy 64:624–635.

Golightly, Jr., R. T., and R. D. Ohmart. 1984. Water economy of two desert canids: Coyote and kit fox. Journal of Mammalogy 65:51–58.

Grinnell, J. 1913. A distributional list of the mammals of California. Proceedings of the California Academy of Science 3:265–390.

Grinnell, J. 1923. A systematic list of the mammals of the California. University of California Publications in Zoology 21:313–324.

Grinnell, J. 1932. Habitat relations of the giant kangaroo rat. Journal of Mammalogy 13:305–320.

Grinnell, J., D. S. Dixon, and J. M. Linsdale. 1937. Fur-bearing mammals of California, Volume 2. Berkeley: University of California Press.

Greenville, A. C., G. M. Wardle, and C. R. Dickman. 2012. Extreme climatic events drive mammal irruptions: Regression analysis of 100-year trends in desert rainfall and temperature. Ecology and Evolution 2:2645–2658.

Hall, E. R. 1946. The mammals of Nevada. Berkeley: University of California Press.

Hall, E. R. 1981. The mammals of North America, 2nd ed. New York: John Wiley and Sons.

Hall, E. R., and K. R. Kelson. 1959. The mammals of North America, Volume 2. New York: Ronald Press.

Hall, F. A., Jr. 1983. Status of the San Joaquin kit fox, *Vulpes macrotis mutica*, at the Bethany wind turbine generating site, Alameda County, California. Sacramento: California Department of Fish and Game.

Harrison, S. W. R., B. L. Cypher, S. Bremner-Harrison, and C. L. Van Horn Job. 2011. Resource use overlap between urban carnivores: Implications for endangered San Joaquin kit foxes (*Vulpes macrotis mutica*). Urban Ecosystems 14:303–311.

Heisey, D. M., and T. K. Fuller. 1985. Evaluation of survival and cause-specific mortality rates using telemetry data. Journal of Wildlife Management 49:668–674.

Hinshaw, J. M. 1997. Noteworthy collections—*Larrea tridentata*. Madroño 44:398–399.

Hoffacker, M. K., M. F. Allen, and R. R. Hernandez. 2017. Land-sparing opportunities for solar energy development in agricultural landscapes: A case study of the Great Central Valley, CA, United States. Environmental Science and Technology 51:14472–14482.

Holland, R. F. 1986. Preliminary descriptions of the terrestrial natural communities of California. Sacramento: California Department of Fish and Game.

HTH (H.T. Harvey and Associates). 2012. California Valley Solar Ranch project, San Luis Obispo County, California: Habitat restoration and revegetation plan with addendum. Fresno, CA: H.T. Harvey and Associates.

HTH (H.T. Harvey and Associates). 2019. California Valley Solar Ranch San Joaquin kit fox monitoring study: Final report. San Luis Obispo, CA: H.T. Harvey and Associates.

Huffman, L., and T. D. Murphy. 1992. The effects of rodenticides and off-road vehicle use on San Joaquin kit fox activity in Bakersfield, California. *In* D. F. Williams, S. Byrne, and T. A. Rado, eds., Endangered and sensitive species of the San Joaquin Valley, California: Their biology, management and conservation, p. 378. Sacramento: California Energy Commission.

Jameson, E. W., Jr., and H. J. Peeters. 1988. California mammals. Berkeley: University of California Press.

Jensen, C. C. 1972. San Joaquin kit fox distribution. Sacramento, CA: Bureau of Sport Fisheries and Wildlife, Division of Wildlife Services.

Johnson, W. E., T. K. Fuller, and W. L. Franklin. 1996. Sympatry in canids: A review and assessment. *In* J. L. Gittleman, ed., Carnivore behavior, ecology, and evolution, Volume 2, pp. 189–218. Ithaca, NY: Cornell University Press.

JSA (Jones and Stokes Associates). 1995. Distribution of the San Joaquin kit fox and effects of military training activities at the Multi-Purpose Range Complex (MPRC) on kit foxes at Fort Hunter Liggett, California—preliminary results. Sacramento, CA: Jones and Stokes Associates.

Kamler, J. F., and W. B. Ballard. 2002. A review of native and nonnative red foxes in North America. Wildlife Society Bulletin 30:370–379.

Kellert, S. R. 1985. Public perceptions of predators, particularly the wolf and coyote. Biological Conservation 31:167–189.

Kelly, P. A., S. E. Phillips, and D. F. Williams. 2005. Documenting ecological change in time and space: The San Joaquin Valley of California. *In* E. A. Lacey and P. Myers, eds. Mammalian diversification: From chromosomes to phylogeography, Publications in Zoology Series, pp. 57–78. Berkeley: University of California Press.

KCPD (Kern County Planning Department). 2014. List of approved/proposed projects for Kern County. Accessed January 12, 2023. https://psbweb.co.kern.ca.us/planning/pdfs/renewable/MasterKernSolarProjectsList.pdf

Kluever, B. M., D. T. Iles, and E. M. Gese. 2019. Ectoparasite burden influences the denning behavior of a small desert carnivore. Ecosphere 10(5): e02749.10.1002/ecs2.2749

Knapp, D. K. 1978. Effects of agricultural development in Kern County, California, on the San Joaquin kit fox in 1977, California Department of Fish and Game, Nongame Wildlife Investigations Final Report, Project E-1-1, Job V-1.21. Sacramento: California Department of Fish and Game.

Koopman, M. E., B. L. Cypher, and J. H. Scrivner. 2000. Dispersal patterns of San Joaquin kit foxes (*Vulpes macrotis mutica*). Journal of Mammalogy 81:213–222.

Koopman, M. E., J. H. Scrivner, and T. T. Kato. 1998. Patterns of den use by San Joaquin kit foxes. Journal of Wildlife Management 62:373–379.

Kurtén, B., and E. Anderson. 1980. Pleistocene mammals of North America. New York: Columbia University Press.

Laughrin, L. L. 1970. San Joaquin kit fox, its distribution and abundance, California Department of Fish and Game, Wildlife Management Branch Administrative Report 70–2. Sacramento: California Department of Fish and Game.

Laurenson, M. K., S. Cleaveland, M. Artois, and R. Woodroffe. 2004. Assessing and managing infectious disease threats to canids. *In* C. Sillero-Zubiri, M. Hoffmann, and D. W. Macdonald, eds., Canids: Foxes, wolves, jackals and dogs: Status survey and conservation action plan, pp. 246–255. Cambridge: IUCN/SSC Canid Specialist Group.

Lewis, J. C., K. L. Sallee, and R. T. Golightly, Jr. 1999. Introduction and range expansion of nonnative red foxes (*Vulpes vulpes*) in California. American Midland Naturalist 142:372–381.

Linnell, J. D. C., and O. Strand. 2000. Interference interactions, coexistence and conservation of mammalian carnivores. Diversity and Distributions 6:169–176.

List, R., and B. L. Cypher. 2004. *Vulpes macrotis. In* C. Sillero-Zubiri, M. Hoffmann, and D. W. Macdonald, eds., Canids: Foxes, wolves, jackals and dogs: Status survey and conservation action plan, pp. 105–109. Cambridge: IUCN/SSC Canid Specialist Group.

Litvaitis, J. A., and D. J. Harrison. 1989. Bobcat-coyote niche relationships during a period of coyote population increase. Canadian Journal of Zoology 67:1180–1188.

Logan, C. G., W. H. Berry, W. G. Standley, and T. T. Kato. 1992. Prey abundance and food habits of San Joaquin kit fox at Camp Roberts Army National Guard Training Site, California, US Department of Energy Topical Report EGG 10617-2158. Springfield, VA: National Technical Information Service.

Loredo, A. I., J. Rudd, J. Foley, D. L. Clifford, and B. L. Cypher. 2020. Climatic suitability of San Joaquin kit fox (*Vulpes macrotis mutica)* dens for sarcoptic mange (*Sarcoptes scabiei*) transmission. Journal of Wildlife Disease 56:126–133.

Macdonald, D. W. 1981. The social organization of the red fox *(Vulpes vulpes). In* J. A. Chapman and D. Pursley, eds., Proceedings of the Worldwide Furbearer Conference, pp. 918–949. Frostburg, MD: Worldwide Furbearer Conference, Inc.

Macdonald, D. W., L. Brown, S. Yerli, and A. Canbolat. 1994. Behavior of red foxes, *Vulpes vulpes*, caching eggs of loggerhead turtles, *Caretta caretta*. Journal of Mammalogy 75:985–988.

Macdonald, D. W., S. Creel, and M. G. Mills. 2004. Canid society. *In* D. W. Macdonald, and C. Sillero-Zubiri, eds., The biology and conservation of wild canids, pp. 85–106. Oxford: Oxford University Press.

MacMahon, J. A., and F. H. Wagner. 1985. The Mojave, Sonoran and Chihuahuan Deserts of North America. *In* M. Evenari, I. Noy-Meir, and D. W. Goodall, eds., Hot deserts and arid shrublands, pp. 105–202. Amsterdam: Elsevier Scientific.

Maldonado, J. E., M. Cotera, E. Geffen, and R. K. Wayne. 1997. Relationships of the endangered Mexican kit fox (*Vulpes macrotis zinseri*) to North American arid-land foxes based on mitochondrial DNA sequence data. Southwestern Naturalist 42:460–470.

McCue, P. M., and T. P. O'Farrell. 1988. Serological survey for selected diseases in the endangered San Joaquin kit fox (*Vulpes macrotis mutica*). Journal of Wildlife Diseases 24:274–281.

McCue, P. M., T. T.Kato, M. L. Sauls, and T. P. O'Farrell. 1981. Inventory of San Joaquin kit fox on land proposed as phase II, Kesterson Reservoir, Merced County, California, US Department of Energy Topical Report EGG 1183–2426. Springfield, VA: National Technical Information Service.

McGee, B. K., W. B. Ballard, K. L. Nicholson, B. L. Cypher, P. R. Lemons, and J. F. Kamler. 2006. Effects of artificial escape dens on swift fox populations in northwest Texas. Wildlife Society Bulletin 34:821–827.

McGinley, M. A. 1984. The adaptive value of male-biased sex ratios among stressed animals. The American Naturalist 124:597–599.

McGrew, J. C. 1979. *Vulpes macrotis*. Mammalian Species 123:1–7.

McMillin, S. C., R. C. Hosea, B. F. Finlayson, B. L. Cypher, and A. Mekebri. 2008. Anticoagulant rodenticide exposure in an urban population of the San Joaquin kit fox. Proceedings of the Vertebrate Pest Conference 23:163–165.

Mercure, A., K. Ralls, K. P. Koepeli, and R. K. Wayne. 1993. Genetic subdivisions among small canids: Mitrochondrial DNA differentiation of swift, kit, and arctic foxes. Evolution 47:1313–1328.

Merriam, C. H. 1888. Description of a new fox from southern California. Proceedings of the Biological Society of Washington 4:5–8.

Merriam, C. H. 1902. Three new foxes of the kit and desert fox groups. Proceedings of the Biological Society of Washington 15:73–74.

Metropolitan Bakersfield Habitat Conservation Plan. 1994. Metropolitan Bakersfield Habitat Conservation Plan. Bakersfield, CA: Metropolitan Bakersfield Habitat Conservation Plan Steering Committee.

Miller, D. S., D. F. Covell, R. G. McLean, W. J. Adrian, M. Niezgoda, J. M. Gustafson, O. J. Rongstad, R. D. Schultz, L. J. Kirk, and T. J. Quan. 2000. Serologic survey for selected infectious disease agents in swift and kit foxes from the western United States. Journal of Wildlife Diseases 36:798–805.

Minnich, R. A. 2008. California's fading wildflowers: Lost legacy and biological invasions. Berkeley: University of California Press.

Minta, S. C., K. A. Minta, and D. F. Lott. 1992. Hunting associations between badgers (*Taxidea taxus*) and coyotes (*Canis latrans*). Journal of Mammalogy 73:814–820.

Moehlman, P. D. 1989. Intraspecific variation in canid social systems. *In* J. L. Git-tleman, ed., Carnivore behavior, ecology, and evolution, pp. 164–182. Ithaca, NY: Cornell University Press.

Moratto, M. J., T. F. King, and W. B. Woolfenden. 1978. Archaeology and California's climate. The Journal of California Anthropology 5:147–161.

Moritz, C. 1994. Defining "evolutionarily significant units" for conservation. Trends in Ecology and Evolution 9:373–375.

Morrell, S. 1971. Life history of the San Joaquin kit fox, California Department of Fish and Game, Federal Aid in Wildlife Restoration Project W-54-R-3 "Special Wildlife Investigations", Final Report. Sacramento: California Department of Fish and Game.

Morrell, S. 1972. Life history of the San Joaquin kit fox. California Fish and Game 58:162–174.

Morrell, S. 1975. San Joaquin kit fox distribution and abundance in 1975, California Department of Fish and Game, Wildlife Management Branch, Administrative Report 75-3. Sacramento: California Department of Fish and Game.

Murdoch, J. D. 2004. Scent marking behavior of the San Joaquin kit fox (*Vulpes macrotis mutica*). Thesis, University of Denver, Denver, CO.

Murdoch, J. D., K. Ralls, B. L. Cypher, and R. P. Reading. 2008a. Barking vocalizations in San Joaquin kit foxes (*Vulpes macrotis mutica*). Southwestern Naturalist 53:118–124.

Murdoch, J. D., K. Ralls, B. L. Cypher, and R. P. Reading. 2008b. Social interactions among San Joaquin kit foxes before, during, and after mating season. Journal of Mammalogy 89:1087–1093.

Nelson, J. L., B. L. Cypher, C. D. Bjurlin, and S. Creel. 2007. Effects of habitat on competition between kit foxes and coyotes. Journal of Wildlife Management 71:1467–1475.

Newsome, S. D., K. Ralls, C. Van Horn Job, M. L. Fogel, and B. L. Cypher. 2010. Stable isotopes evaluate exploitation of anthropogenic foods by the endangered San Joaquin kit fox *(Vulpes macrotis mutica)*. Journal of Mammalogy 91:1313–1321.

Niedringhaus, K. D., J. D. Brown, K. M. Sweeley, and M. J. Yabsley. 2019. A review of sarcoptic mange in North American Wildlife. International Journal of Parasites: Parasites and Wildlife 9:285–297.

O'Farrell, T. P. 1987. Kit fox. *In* M. Novak, J. A. Baker, M. E. Obbard, and B. Malloch, eds., Wild furbearer management and conservation in North America, pp. 423–431. Toronto, Ontario: Ontario Trappers Association and Ontario Ministry of Natural Resources.

O'Farrell, T. P., W. H. Berry, and G. D. Warrick. 1987. Distribution and status of the endangered San Joaquin kit fox, *Vulpes macrotis mutica*, on Fort Hunter Liggett and Camp Roberts, California, US Department of Energy Topical Report EGG 10282-2194. Springfield, VA: National Technical Information Service.

O'Farrell, T. P., and L. Gilbertson. 1979. Ecology of the desert kit fox, *Vulpes macrotis arsipus,* in the Mojave Desert of Southern California. Bulletin of the Southern California Academy of Science 85:1–15.

O'Farrell, T. P., T. T. Kato, P. M. McCue, and M. L. Sauls. 1980. Inventory of San Joaquin kit fox on USBLM lands in southern and southwestern San Joaquin Valley, US Department of Energy Topical Report EGG 1183-2400. Springfield, VA: National Technical Information Service.

Oftedal, O. T., and J. L. Gittleman. 1989. Patterns of energy output during reproduction in carnivores. *In* J. L. Gittleman, ed., Carnivore behavior, ecology, and evolution, pp. 355–378. Ithaca, NY: Cornell University Press.

Orloff, S. G., F. Hall, and L. Spiegel. 1986. Distribution and habitat requirements of the San Joaquin kit fox in the northern extreme of their range. Transactions of the Western Section of The Wildlife Society 22:60–70.

Ostfeld, R. S., and F. Keesing. 2000. Pulsed resources and community dynamics of consumers in terrestrial ecosystems. Trends in Ecology and Evolution 15:232–237.

Packard, R. L., and J. H. Bowers. 1970. Distributional notes on some foxes from western Texas and eastern New Mexico. Southwestern Naturalist 14:450–451.

Peabody, F. E., and J. M. Savage. 1958. Evolution of a Coast Range corridor in California and its effects on the origin and dispersion of living amphibians and reptiles. American Association for the Advancement of Science, Publication 51:159–186.

Pearce, D., J. Strittholt, T. Watt, and E. N. Elkind. 2016. A path forward: Identifying least-conflict solar PV development in California's San Joaquin Valley. Corvallis, OR: Conservation Biology Institute.

Peek, J., B. Dale, H. Hristienko, L. Kantar, K. A. Loyd, S. Mahoney, C. Miller, D. Murray, L. Oliver, and C. Soulliere. 2012. Management of large mammalian carnivores in North America, Technical Review 12–01. Bethesda, MD: The Wildlife Society.

Pence, D. B., and E. Ueckermann. 2002. Sarcoptic mange in wildlife. Review of Science and Technology 21:385–398.

Phillips, S. E., and B. L. Cypher. 2019. Solar energy development and endangered species in the San Joaquin Valley, CA: Identification of conflict zones. Western Wildlife 6:29–44.

Pianka, E. R. 1978. Evolutionary ecology, 2nd ed. New York: Harper and Row Publishers.

Polis, G. A., C. A. Myers, and R. D. Holt. 1989. The ecology and evolution of intraguild predation: Potential competitors that eat each other. Annual Review of Ecology and Systematics 20:297–330.

Previtali, M. A., M. Lima, P. L. Meserve, D. A. Kelt, and J. R. Gutiérrez. 2009. Population dynamics of two sympatric rodents in a variable environment: Rainfall, resource availability, and predation. Ecology 90:1996–2006.

Prugh, L. R., N. Deguines, J. B. Grinath, K. N. Suding, W. T. Bean, R. Stafford, and J. S. Brashares. 2018. Ecological winners and losers of extreme drought in California. Nature Climate Change Letters 8:819–824.

PPIC (Public Policy Institute of California). 2006. Just the facts: California's Central Valley. June. Accessed January 12, 2023. https://www.ppic.org/content/pubs/jtf/JTF_CentralValleyJTF.pdf

PPIC (Public Policy Institute of California). 2021. California Poverty by County and Legislative District. Accessed January 12, 2023. https://www.ppic.org/interactive/california-poverty-by-county-and-legislative-district/

Ralls, K., B. L. Cypher, and L. K. Spiegel. 2007. Social monogamy in kit foxes: Formation, association, duration, and dissolution of mated pairs. Journal of Mammalogy 88:1439–1446.

Ralls, K., K. Pilgrim, P. J. White, E. E. Paxinos, and R. C. Fleischer. 2001. Kinship, social relationships and den use in kit foxes. Journal of Mammalogy 82:858–866.

Ralls, K., and D. A. Smith. 2004. Latrine use by San Joaquin kit foxes (*Vulpes mac-*

rotis mutica) and coyotes (*Canis latrans*). Western North American Naturalist 64:544–547.

Ralls, K., and P. J. White. 1995. Predation on San Joaquin kit foxes by larger canids. Journal of Mammalogy 76:723–729.

Ralls, K., and P. J. White. 2003. Diurnal spacing patterns in kit foxes, a monogamous canid. Southwestern Naturalist 48:432–436.

Reese, E. A., W. G. Standley, and W. H. Berry. 1992. Habitat, soils, and den use of San Joaquin kit fox (*Vulpes velox macrotis*) at Camp Roberts Army National Guard Training Site, California, US Department of Energy Topical Report EGG 10617-2156. Springfield, VA: National Technical Information Service.

Reichman, O. J., and M. V. Price. 1993. Ecological aspects of heteromyid foraging. *In* H. H. Genoways, and J. H. Brown, eds. Biology of the Heteromyidae, American Society of Mammalogists, Special Publication No. 10, pp. 618–651. Provo, UT: Brigham Young University.

REN21.2022. Renewables 2022 global status report. Accessed January 12, 2023. https://www.ren21.net/wp-content/uploads/2019/05/GSR2022_Full_Report.pdf

Riner, A. J., J. L. Rudd, D. L. Clifford, B. L. Cypher, J. E. Foley, and P. Foley. 2018. Comparison of flea (Siphonaptera) burdens on the endangered San Joaquin kit fox (*Vulpes macrotis mutica* (Carnivora, Canidae)) inhabiting urban and non-urban environments in Central Valley, California. Journal of Medical Entomology 55:995–1001.

del Rio, C. M., B. Dugelby, D. Foreman, B. Miller, R. Noss, and M. Phillips. 2001. The importance of large carnivores to healthy ecosystems. Endangered Species Update 18:202–215.

Robinson, W. B. 1961. Population changes of carnivores in some coyote control areas. Journal of Mammalogy 42:510–515.

Rohwer, S. A., and D. L. Kilgore, Jr. 1973. Interbreeding in the arid-land foxes, *Vulpes velox* and *V. macrotis*. Systematic Zoology 22:157–165.

Sacks, B. N., M. J. Statham, J. D. Perrine, S. M. Wisely, and K. A. Aubry. 2010. North American montane red foxes: Expansion, fragmentation, and the origin of the Sacramento Valley red fox. Conservation Genetics 11:1523–1539.

Savage, D. E., and D. E. Russell. 1983. Mammalian paleofaunas of the world. Menlo Park, CA: Benjamin/Cummings.

Sawyer, J. O., T. Keeler-Wolf, and J. M. Evens. 2009. A manual of California vegetation. Second edition. Sacramento: California Native Plant Society.

Schitoskey, F., Jr. 1975. Primary and secondary hazards of three rodenticides to kit fox. Journal of Wildlife Management 19:416–418.

Schwartz, M. K., K. Ralls, D. F. Williams, B. L. Cypher, K. L. Pilgrim, and R. C. Fleischer. 2005. Gene flow among San Joaquin kit fox populations in a severely changed ecosystem. Conservation Genetics 6:25–37.

Scott, J. M., D. D. Goble, J. A. Wiens, D. S. Wilcove, M. Bean, and T. Male. 2005. Recovery of imperiled species under the Endangered Species Act: The need for a new approach. Frontiers in Ecology and the Environment 3:383–389.

Scrivner, J. H., T. P. O'Farrell, K. Hammer, and B. L. Cypher. 2016. Translocation of the endangered San Joaquin kit fox, *Vulpes macrotis mutica*: A retrospective assessment. Western North American Naturalist 76:90–100.

Scrivner, J. H., T. P. O'Farrell, and T. T. Kato. 1987. Dispersal of San Joaquin kit foxes, *Vulpes macrotis mutica*, on Naval Petroleum Reserve #1, Kern County, California, US Department of Energy Topical Report EGG 10282-2190. Springfield, VA: National Technical Information Service.

Seton, E. T. 1929. Lives of game animals. Garden City, NY: Doubleday.

Shaw, W. T. 1934. The ability of the giant kangaroo rat as a harvester and storer of seeds.Journal of Mammalogy 15:275–287.

Sillero-Zubiri, C., M. Hoffmann, and D. W. Macdonald, eds. 2004. Canids: Foxes, wolves, jackals, and dogs: Status survey and conservation action plan. Cambridge: IUCN/SSC Canid Specialist Group.

Single, J. R., D. J. Germano, and M. H. Wolfe. 1996. Decline of kangaroo rats during a wet winter in the southern San Joaquin Valley, California. Transactions of the Western Section of The Wildlife Society 32:34–41.

Smith, D. A., K. Ralls, B. L. Cypher, H. O. Clark, Jr., P. A. Kelly, D. F. Williams, and J. E. Maldonado. 2006. Relative abundance of endangered San Joaquin kit foxes (*Vulpes macrotis mutica*) based on scat-detection dog surveys. Southwestern Naturalist 51:210–219.

Smith, D. A., K. Ralls, B. L. Cypher, and J. E. Maldonado. 2005. Assessment of scat-detection dog surveys to determine kit fox distribution. Wildlife Society Bulletin 33:897–904.

SEIA (Solar Energy Industries Association). 2016. Solar industry research data. Accessed January 12, 2023. https://www.seia.org/research-resources/solar-industry -data

Spencer, K. A., W. H. Berry, W. G. Standley, and T. P. O'Farrell. 1992. Reproduction of the San Joaquin kit fox on Camp Roberts Army National Guard Training Site, California, US Department of Energy Topical Report EGG 10617-2154. Springfield, VA: National Technical Information Service.

Spencer, K. A., and H. J. Egoscue. 1992. Fleas of the San Joaquin kit fox (*Vulpes macrotis mutica*) on Camp Roberts Army National Guard Training Site, California, US Department of Energy Topical Report EGG 10617-2161. Springfield, VA: National Technical Information Service.

Spiegel, L. K., ed. 1996. Studies of San Joaquin kit fox in undeveloped and oil-developed areas. Sacramento: California Energy Commission.

Spiegel, L. K., and M. Bradbury. 1992. Home range characteristics of the San Joaquin kit fox in western Kern County, California. Transactions of the Western Section of The Wildlife Society 28:83–92.

Sproul, M. J., and M. A. Flett. 1993. Status of San Joaquin kit fox in the northwest margin of its range. Transactions of the Western Section of the Wildlife Society 29:61–99.

Standley, W. G., W. H. Berry, T. P. O'Farrell, and T. T. Kato. 1992. Mortality of San Joaquin kit fox at Camp Roberts Army National Guard Training Site, California, US Department of Energy Topical Report EGG 10627-2157. Springfield, VA: National Technical Information Service.

Standley, W. G., and P. M. McCue. 1997. Prevalence of antibodies against selected diseases in San Joaquin kit foxes at Camp Roberts, California. California Fish and Game 83:30–37.

Stephens, D. W., J. S. Brown, and R. C. Ydenberg. 2007. Foraging behavior and ecology. Chicago: University of Chicago Press.

Stromberg, M. R., and M. S. Boyce. 1986. Systematics and conservation of the swift fox, *Vulpes velox*, in North America. Biological Conservation 35:97–110.

Swaisgood, R. R., J.-P. Montagne, C. M. Lenihan, C. L. Wisinski, L. A. Nordstrom, and D. M. Shier. 2019. Capturing pests and releasing ecosystem engineers: Translocation of common but diminished species to re-establish ecological roles. Animal Conservation 22:600–610.

Tennant, E. N., and D. J. Germano. 2013. Competitive interactions between Tipton and Heermann's kangaroo rats in the San Joaquin Valley, California. The Southwestern Naturalist 58:258–264.

Tennant, E. N., D. J. Germano, and B. L. Cypher. 2013. Translocating endangered kangaroo rats in the San Joaquin Valley of California: Recommendations for future efforts. California Fish and Game 99:90–103.

Thornton, W. A., and G. C. Creel. 1975. The taxonomic status of kit foxes. Texas Journal of Science 26:127–136.

USBLM (US Bureau of Land Management). 2010. Carrizo Plain National Monument Approved Resource Management Plan and Record of Decision. Bakersfield, CA: US Department of Interior, Bureau of Land Management.

USCB (US Census Bureau). 2019. Population and housing unit estimates tables. Accessed January 12, 2023. https://www.census.gov/programs-surveys/popest/data/tables.2019.html

USDA (US Department of Agriculture). 1997. Environmental Assessment: Wildlife damage management for the protection of livestock, property and human health and safety in the California ADC South and San Luis Districts. Sacramento, CA: US Department of Agriculture, Animal and Plant Health Inspection Service, Animal Damage Control.

USFWS (US Fish and Wildlife Service). 1967. Native fish and wildlife. Endangered species. Federal Register 32:4001.

USFWS (US Fish and Wildlife Service). 1983. San Joaquin kit fox recovery plan. Portland, OR: US Fish and Wildlife Service, Region 1.

USFWS (US Fish and Wildlife Service). 1998. Recovery plan for upland species of the San Joaquin Valley, California. Portland, OR: US Fish and Wildlife Service, Region 1.

USFWS (US Fish and Wildlife Service). 2010. San Joaquin kit fox 5-year review: Summary and evaluation. Portland, OR: US Fish and Wildlife Service, Region 1.

USFWS (US Fish and Wildlife Service). 2020a. San Joaquin kit fox 5-year review: Summary and evaluation. Portland, OR: US Fish and Wildlife Service, Region 1.

USFWS (US Fish and Wildlife Service). 2020b. Species status assessment report for the San Joaquin kit fox (*Vulpes macrotis mutica*). Sacramento, CA: US Fish and Wildlife Service.

USGS (US Geological Survey). 2021. California Water Science Center, California's Central Valley. Accessed January 12, 2023. https://ca.water.usgs.gov/projects/central-valley/about-central-valley.html

Visit Bakersfield. 2021. Bakersfield facts and information. Accessed January 12, 2023. https://www.visitbakersfield.com/media/facts-and-information/

Waithman, J., and A. Roest. 1977. A taxonomic study of the kit fox, *Vulpes macrotis*. Journal of Mammalogy 58:157–164.

Warrick, G. D., H. O. Clark, Jr., P. A. Kelly, D. F. Williams, and B. L. Cypher. 2007. Use of agricultural lands by San Joaquin kit foxes. Western North American Naturalist 67:270–277.

Warrick, G. D., and B. L. Cypher. 1998. Factors affecting the spatial distribution of a kit fox population. Journal of Wildlife Management 62:707–717.

Warrick, G. D., and B. L. Cypher. 1999. Variation in body mass of San Joaquin kit foxes. Journal of Mammalogy 80:972–979.

Westall, T. L., B. L. Cypher, K. Ralls, and D. J. Germano. 2019. Raising pups of urban San Joaquin kit fox: Relative roles of adult group members. Western North American Naturalist 79:364–377.

Westall, T. L., B. L. Cypher, K. Ralls, and T. Wilbert. 2018. Observations of social polygyny, allonursing, extrapair copulation, and inbreeding in urban San Joaquin kit foxes (*Vulpes macrotis mutica*). Southwestern Naturalist 63:271–276.

WRCC (Western Regional Climate Center). 2021. Annual precipitation map for western United States. Accessed January 12, 2023. https://wrcc.dri.edu/Climate/precip_map_show.php?map=0

White, P. J., W. H. Berry, J. J. Eliason, and M. T. Hanson. 2000. Catastrophic decrease in an isolated population of kit foxes. Southwestern Naturalist 45:204–211.

White, G. C., and R. A. Garrott. 1990. Analysis of wildlife radiotracking data. San Diego, CA: Academic Press.

White, P. J., and R. A. Garrott. 1997. Factors regulating kit fox populations. Canadian Journal of Zoology 75:1982–1988.

White, P. J., and R. A. Garrott. 1999. Population dynamics of kit foxes. Canadian Journal of Zoology 77:486–493.

White, P. J., and K. Ralls. 1993. Reproduction and spacing patterns of kit foxes relative to changing prey availability. Journal of Wildlife Management 57:861–867.

White, P. J., K. Ralls, and C. A. Vanderbilt White. 1995. Overlap in habitat and food use between coyotes and San Joaquin kit foxes. Southwestern Naturalist 40:342–349.

White, P. J., C. A. Vanderbilt White, and K. Ralls. 1996. Functional and numerical responses of kit foxes to a short-term decline in mammalian prey. Journal of Mammalogy 77:370–376.

Wilbert, T. R. 2013. Patterns and processes of genetic diversity in the endangered San Joaquin kit fox. Dissertation, George Mason University, Fairfax, VA.

Wilbert, T. R., J. E. Maldonado, M. T. N. Tsuchiya, M. Sikaroodi, B. L. Cypher, C. Van Horn Job, K. Ralls, and P. M. Gillett. 2020. Patterns of MHC polymorphism in endangered San Joaquin kit foxes living in urban and non-urban environments. *In* J. Ortega and J. E. Maldonado, eds., Conservation genetics in mammals: Integrative research using novel approaches, pp. 269–298. New York: Springer.

Williams, D. F. 1992. Geographic distribution and population status of the giant kangaroo rat, *Dipodomys ingens* (Rodentia, Heteromyidae). *In* D. F. Williams, S. Byrne, and T. A. Rado, eds., Endangered and sensitive species of the San Joaquin Valley, California: Their biology, management and conservation, pp. 301–328. Sacramento: California Energy Commission.

Williams, D. F., and D. J. Germano. 1992. Recovery of endangered kangaroo rats in the San Joaquin Valley, California. Transactions of the Western Section of The Wildlife Society 28:93–106.

Wilson, D. S., A. B. Clark, K. Coleman, and T. Dearsyne. 1994. Shyness and boldness in humans and other animals. Trends in Ecology and Evolution 9:442–446.

World Population Review. 2021. Bakersfield. Accessed January 12, 2023. https://worldpopulationreview.com/us-cities/bakersfield-ca-population

Zeh, J. A., and D. W. Zeh. 2001. Reproductive mode and the genetic benefits of polyandry. Animal Behavior 61:1051–1063.

Zoellick, B. W., C. E. Harris, B. T. Kelly, T. P. O'Farrell, T. T. Kato, and M. E. Koopman. 2002. Movements and home ranges of San Joaquin kit foxes relative to oilfield development. Western North American Naturalist 62:151–159.

Zoellick, B. W., T. P. O'Farrell, P. M. McCue, C. E. Harris, and T. T. Kato. 1987. Reproduction of the San Joaquin kit fox on Naval Petroleum Reserve #1, Elk Hills, California, 1980–1985, US Department of Energy Topical Report EGG 10282-2144. Springfield, VA: National Technical Information Service.

INDEX

activity, 89
adoption, 106–107
After the Grizzly (Alagona), 176, 177*fig*, 182
agricultural lands, 31, 86, 131–136, 132*fig*, 134*fig*, 164–165, 167
Alagona, Peter, 176, 177*fig*, 182
Alameda County, 167
Alderton, D., 16
Alkali Meadow, 34
Alkali Playa, 34
Allensworth Ecological Reserve, 35, 39
allonursing, 106
Altithermal climatic age, 23
ambassador animals, 187, 188*fig*
Anathermal climatic age, 23
animal movement corridors, 124
Annual Grassland, 34
anthropogenic food, 145
anthropogenic habitats, 85
Arctic foxes, 112

badgers, 48, 90, 91–92, 112, 116–117
Bakersfield
    conservation efforts and, 180
    description of, 140
    habitat in, 30, 148–149
    history of kit foxes in, 140–142
    human interactions in, 152–153
    images of kit foxes in, 118*fig*, 141*fig*, 146*fig*, 147*fig*, 150*fig*
    interspecific interactions in, 116–118, 118*fig*

    population density in, 145
    population size in, 154, 156
    research potential in, 156–157, 165
    resource abundance and, 81, 85
    sarcoptic mange in, 54, 56, 58
    survival rates and, 45
    Taung Ming-Lin incident and, 178
Bakersfield College (BC), 107*fig*, 178, 179*fig*, 180
Balestreri, Antonio, 31, 166–167
Beedy, E. C., 142
behavior, 108–112
Bena Landfill, 77
Biological Opinion or Habitat Conservation Plan, 182
Blackwells Corner Solar Farm, 175
bobcats, 22–23, 40, 45, 46*fig*, 47, 148
body form, 7
Bowers, J. H., 17
breeding/mating, 13, 87, 103–104. *See also* reproduction
Bremner-Harrison, S., 112
Briden, Laurie, 167
Buena Vista Valley, 35
burrowing owls, 101–102, 117, 136

caching food, 110, 112
Caliente Range, 29, 184
California Department of Fish and Wildlife, 18, 26, 56, 71, 123, 165, 167, 173, 178, 181
California Endangered Species Act (1972), 2, 14, 19, 170, 171

California Energy Commission (CEC), 121, 165, 173
California Fish and Game Commission, 170
California Flats Solar Plant, 129, 175
California High Speed Rail Project, 175
California Living Museum (CALM), 56–57, 167, 187, 188*fig*
California State University-Bakersfield (CSUB), 116*fig*, 178, 179*fig*, 180
California State University-Stanislaus (CSUS), 167
California Valley Solar Ranch (CVSR), 80, 81, 126–127, 175
Camp Roberts area, 30–31, 35, 48, 50, 53, 73, 77, 86, 166–167
canids, early, 16
carnivores, in fossil record, 16
Carrizo Plain
    conservation efforts and, 165, 181, 184
    drought conditions in, 72–73, 77
    home ranges and, 81
    photograph of, 37*fig*
    population in, 26–28, 29
    prey availability in, 63, 68, 69, 75, 75*fig*, 78
    San Joaquin Desert and, 24
    solar fields/farms and, 126–127
    survival and mortality rates in, 41, 48, 49
Carrizo Plain National Monument, 26, 35, 173, 174*fig*, 181
Catalina Island, 54
cats, 115
Center for Natural Lands Management, 27
cestodes, 53
Chevron, 166
Cholame Valley area, 28
Clevenger, A. P., 139
climate change, 158–161, 160*fig*
climatic ages, 23
Clinton, Bill, 173
Coalinga, 140, 157
Coast Range Saltbush Scrub, 34
Coles Levee Ecopreserve, 120
coloring, 5

concrete dens, 98*fig*, 99*fig*
conservation efforts, 124, 126
conservation movement, 176
coyotes
    agricultural lands and, 86
    badgers and, 112
    comparison with, 7
    dens and, 90
    images of, 8*fig*
    misidentification and, 7, 33, 51
    population dynamics and, 72
    predation risk and, 13, 22–23, 40, 45, 46*fig*, 113–114, 148, 152, 165
    predator control programs and, 31, 168
    red foxes and, 48, 115
    sarcoptic mange and, 58–59
    solar fields/farms and, 127
crossing structures, 138–139
Cuyama Valley, 24, 29, 184
Cypher, B. L., 10, 34, 90, 97, 129, 142, 165, 167, 169

Dash, 188*fig*
den switching, 52–53, 62
dens
    artificial, 95, 97, 98*fig*, 100*fig*, 124, 136
    atypical, 95
    characteristics of, 91–97
    earthen, 12, 12*fig*
    ejecta and, 93–94
    entrances of, 92*fig*, 92*t*
    images of, 92*fig*, 99*fig*, 100*fig*
    material excavated from, 93*fig*
    natal, 86, 94–95, 95*fig*, 96*fig*, 97, 100, 154
    other animals entering, 116*fig*, 118*fig*
    overlapping use of, 59
    overview of, 89–91
    soil types and, 40, 91
    survey efforts and, 164
    urban areas and, 149, 153–154
    use patterns of, 97–102, 101*t*
dental formula, 10
desert kit foxes, 14
dew claws, 7
diet, 12–13, 22, 72–73, 74–79, 110, 145, 148

disease, 31–32, 53–54, 152, 191
dispersal, 88–89, 108
distemper, 54, 152, 191
distribution, 21–33, 27*fig*, 29*fig*
dogs, 48, 86, 115–116
Dragoo, J. W., 16, 18

earthen dens, 12, 12*fig*
ecological plasticity, 22, 34, 81, 122
Ecological Reserves, 173
EG&G Energy Measurements, Inc.,
    165
Egoscue, H. J., 17, 76, 109
Elk Hills, 49, 68, 69–70, 73, 77, 81, 120,
    121, 174
Elkhorn Plain, 35, 70
Endangered Species Act (1973), 14, 18,
    19, 120, 171, 178
Endangered Species Preservation Act
    (1966), 170
Endangered Species Recovery Program
    (ESRP), 10, 56, 142–143, 167
endangered status, 14
endowments, 181–182
entombment, 51, 133, 165
Entrix, Inc., 159
evaporative heat loss, 12
Evolutionarily Significant Units (ESUs),
    19
exploitative competition, 22, 113–114,
    115
extra-pair copulations, 13, 86, 87,
    103–104, 109

fecal deposits, 110, 111*fig*
fencing, permeable security, 123, 126
fleas, 51–53, 52*fig*, 62, 100, 153
foraging ecology, 73–79
Fort Hunter Liggett, 41, 86, 166

Garrott, R. A., 72
genetic exchange, 15
Germano, D. J., 25–26
gestation, 60, 63–64
golden eagles, 47–48, 47*fig*, 113, 127
Golightly, R. T., Jr., 11
gophers, 90, 136
GPS-enabled collars, 87

gray foxes, 7, 8–9*fig*, 54, 90, 114
grazing lands, 134, 134*fig*
great-horned owls, 48
Grinnell, J., 1–2, 7, 10, 17–18, 31, 32, 33,
    163–164, 168, 180
ground cover, 38–39
groundwater, 135
guilds, 113

habitat conservation plans (HCPs), 174,
    182
habitat loss and degradation, 2, 13–14,
    121–122, 132–133, 140, 168–169, 171,
    189–190
habitat restoration, 135–136
habitat suitability, 27*fig*, 29*fig*, 33–41,
    36*fig*, 37*fig*, 38*fig*
habitat use, 85–86
habitats, modified, 45
Hall, E. R., 17, 18
Hall, F., 167
HDPE dens, 97, 98*fig*, 99*fig*, 100*fig*
"helpers," 104–106
Hills/Blackwell Solar Farm, 131
Holocene, 23
home ranges, 79–85, 82–83*t*, 84*fig*, 148,
    169
humidity, 90
hunting behavior, 110, 112
hybridization, 17

Incidental Take Permit, 182
infectious pathogens, 53
interference competition, 22, 45, 48,
    113
interspecific interactions, 113–118
intraguild competition, 113, 114
intraspecific aggression, lack of evi-
    dence for, 85, 109–110
invertebrates
    den use by, 118
    habitat restoration and, 136
    as prey, 72–73, 77–78, 79
island foxes, 54
IUCN, 18–19

jackrabbits, 69, 76, 136
Jensen, C. C., 134, 140, 164

kangaroo rats
  agricultural lands and, 134, 136
  availability of, 61, 70–71, 164, 166
  burrow systems of, 75*fig*
  decline in number of, 171
  dens and, 90, 91
  habitat and, 38, 39–40
  home ranges and, 80–81
  hunting behavior and, 75–76, 78–79,
    110
  images of, 61*fig*, 68*fig*, 74*fig*
  northern range and, 32
  rainfall and, 65, 68–69
  San Joaquin Desert and, 24, 25
  solar fields/farms and, 127
Kato, T., 165
Kelly, P., 167, 180
Kelson, K. R., 18
Kern County, 26–28, 35, 41, 47, 48, 49,
    69, 71, 81, 88, 133, 164, 174, 175, 184
Kern Front, 30
Kern National Wildlife Refuge, 28,
    29–30
Kern River corridor, 142, 148
Kettleman Hills, 28
Kilgore, D. L., Jr., 17
kleptoparasitism, 116
Kluever, B. M., 53
Knapp, D., 51, 133, 164–165
Koopman, M. E., 88

lactation, 64, 104
latrines, 110
Laughrin, L., 33–34, 164, 169
Lewis and Clark expedition, 16
litter size, 64, 145
lizards, 24, 25, 65, 78, 118, 136, 171
Lokern Natural Area, 35, 37*fig*, 48, 49,
    70, 73, 81, 120
Loredo, A. I., 59, 90

mange, 54, 55*fig*, 56–60, 56*t*, 57*fig*, 157,
    191
mating/breeding, 13, 87, 103–104. *See
    also* reproduction
McCue, P. M., 54
measurements, 10
Medithermal climatic age, 23

Merced County, 28, 32, 35, 41, 49, 77,
    167, 175, 184–185
Mercure, A., 167
metabolic water, 11
metapopulation viability analysis
    (MVA), 185–186
Metropolitan Bakersfield Habitat Con-
    servation Plan (MBHCP), 174
mice, 77
Ming-Lin, 177–178
minimum convex polygon (MCP)
    method, 80
Mojave Desert species, 23, 24
monogamy, 102–103
Monterey County, 28, 30, 53, 166
Moratto, M. J., 23
Morrell, S., 61, 93, 164, 169
mortality sources, 13, 45–51, 46*fig*, 47*fig*,
    49*fig*, 113, 115–116, 133, 136–137,
    143, 171
movements, 86–87
Murdoch, J. D., 90, 97, 109
Murphy, T., 140, 165

Nature Conservancy, The, 173
Naval Petroleum Reserves, 45, 63, 71,
    72, 88, 121, 165–166, 167
nematodes, 53
netting, danger of, 143
nocturnal activity patterns, 12, 13, 89,
    90
non-native grasses/plants, 38–39, 68, 69
Northern Carrizo Ecological Reserve,
    84*fig*
Northern Claypan Vernal Pool, 34
Northern Hardpan Vernal Pool, 34
northern range, 32–33
Northern Semitropic Ridge Ecological
    Reserve, 84*fig*
Nuevo-torch, 120

obligatory monogamy, 102
O'Farrell, T., 54, 165
Ohmart, R. D., 11
oil fields, 45, 50–51, 119–123, 121*fig*,
    122*t*, 165–166
orchards, 86
outreach efforts, 186–188

Packard, R. J., 17
Panoche Solar Project, 175
Panoche Valley region, 27–28, 35, 41, 54, 133, 167, 184
Panoche Valley Solar Farm (PVSR), 126–127
panting, 11
parasites, 51–53, 52*fig*, 54–60, 55*fig*, 56*t*, 57*fig*
passive heat dissipation, 12
pelage, 5
perennial monogamy, 102
permeable security fencing, 123, 126
Phillips, S. E., 129
Pixley National Wildlife Refuge, 39
Pleistocene, 23
population dynamics, 64–73, 70*t*
population estimates, 168–170
population viability analyses (PVAs), 185–186
precipitation, 10, 39, 64–65, 68–70, 70*t*, 71–72
predation risk, 22–23, 31, 45, 133, 166
predator control programs, 31, 168, 170–171
predators
    attention paid to, 1
    role of, 3
prey availability, 41, 63, 65, 68–72, 80–81, 132–133, 135–136, 166
protected fur-bearer status, 170

rabbits, 40, 65, 76–77
rabies, 31–32, 53–54, 117, 152
raccoons, 152
rainfall, 10, 39, 64–65, 68–70, 70*t*, 71–72
Ralls, Katherine, 103, 165, 167
range map, 15*fig*
rattlesnakes, 118
Recovery Plan for Upland Species of the San Joaquin Valley, California, 171–172, 172*fig*, 174
red foxes, 7, 8*fig*, 16, 31, 48, 54, 58, 86, 90, 102, 112, 114–115, 116*fig*, 152
reproduction, 60–64, 66–67*t*, 143, 145
Resource Dispersion Hypothesis, 105–106

roads, 48–49, 49*fig*, 136–140, 137*fig*, 139*fig*
rodenticides, 13, 50, 133, 134, 143
Rohwer, S. A., 17
roundworms, 53

Salinas Valley, 30–32, 53, 85, 166–167
San Joaquin Desert, 2, 3, 24–26, 25*fig*, 38
San Joaquin kit fox
    activity of, 89
    adaptations of, 11–12
    agricultural lands and, 86, 131–136, 164–165, 167, 190–191
    behavior of, 108–112
    biopolitics and, 176–183
    climate change and, 158–161, 190, 191
    conservation efforts regarding, 168–176, 191
    dens of. *See* dens
    description of, 5–10
    dispersal of, 88–89
    distribution and habitat of, 21–40
    ecology of, 12–14
    evolution and taxonomy of, 15–19
    foraging ecology and, 73–79
    future of, 189–193
    habitat use and, 85–86
    home range of, 79–85
    images of, 6*fig*, 12*fig*
    interspecific interactions and, 113–118
    movements of, 86–87
    oil fields and, 45, 50–51, 119–123, 165–166
    population dynamics of, 64–73
    pups, 47–48, 61–63, 61*fig*, 62*fig*, 63*fig*, 79, 89, 104–107, 105*fig*, 107*fig*, 192*fig*
    range map for, 15*fig*
    reproduction of, 60–64
    research and conservation needs regarding, 183–188
    research efforts regarding, 163–168
    roads and, 136–140
    social ecology of, 102–108
    solar fields and, 45, 123–131, 175, 190
    status of, 14–15
    survival and mortality factors of, 41–60, 72

San Joaquin kit fox (*continued*)
    urban areas and, 30, 45, 102, 106,
        140–158, 189–190, 191–192
San Luis Obispo County, 26, 27, 35, 47,
    53, 77, 166, 173, 175, 184
San Luis Reservoir, 35
sarcoptic mange, 54, 55*fig*, 56–60, 56*t*,
    57*fig*, 157, 191
satellite populations, 28–29, 30, 41, 73,
    167, 184–185
Say, T., 16
scent-marking, 85, 86, 110
Scott, J. M., 192
selamectin, 56–57
Semitropic Ridge area, 81, 85, 133
Seton, 17
shy/bold behavior, 112
skunks, 31–32, 53–54, 102, 117, 152
Smithsonian Institution, 165
social ecology, 102–108
social units, 13
soil types, 40, 91
solar fields/farms, 26, 45, 123–131,
    125*fig*, 128*fig*, 129*fig*, 130*t*, 131*t*, 175
species status assessments, 172
Spiegel, L., 10, 48, 51, 73, 165
squirrels, 77, 89, 90, 91–92, 102, 118,
    134, 136
subcaudal glands, 7
survival rates, 41–45, 42–44*t*, 72, 143
Sustainable Groundwater Management
    Act (SGMA), 190–191
Sustainable Groundwater Management
    Act (SGMA; 2014), 135, 161
swift foxes, 7, 16–17, 18–19, 97, 112

Taft, 59, 140, 157
Taft College (TC), 178, 179*fig*, 180
tapeworms, 53
Taung Ming-Lin incident, 177–178
teeth, 10
Temblor Range, 26
temperature/heat tolerance, 11–12, 90
thermoregulation, 11–12, 13

ticks, 52
toes, 7
Topaz Solar Farms (TSF), 80, 81, 126–
    127, 129*fig*, 175
toxins, 50, 133, 143, 168, 171

Udall, Stewart, 170
umbrella species, 171–172
Upper Sonoran Subshrub Scrub, 34
urban areas, 30, 45, 102, 106, 140–158,
    141*fig*, 144*fig*, 146*fig*, 147*fig*, 150–
    151*fig*, 155*fig*
US Army, 166
US Bureau of Land Management (BLM),
    26, 27, 173, 181
US Department of Energy, 165
US Fish and Wildlife Service (USFWS),
    18, 123, 159, 171–172, 176, 177–178,
    181

Valley Oak Woodland, 34
Valley Saltbush Scrub, 34
Valley Sink Scrub, 34
vegetation management, 123–124, 135
vehicle strikes, 48–49, 49*fig*, 137, 137*fig*,
    139*fig*, 143
vocalizations, 108–109

Warrick, G. D., 10, 133, 142
water availability, 190–191
water/water loss, 11
Wayne, R., 18, 167
weights, 10
Westervelt Ecological Services, 185
White, G. C., 72
White, P. J., 73, 165
Wilbert, T. R., 156
wild oats, 39
Wildlands Conservancy, The (TWC), 183
Williams, D., 167
Wind Wolves Preserve, 183
Wright Solar Plant/Project, 131, 175, 185

Zoellick, B. W., 63–64, 86